NUREG/CR-6931, Vol. 3
NISTIR 7472

Cable Response to Live Fire (CAROLFIRE) Volume 3: Thermally-Induced Electrical Failure (THIEF) Model

Manuscript Completed: January 2008
Date Published: April 2008

Prepared by
K. McGrattan

National Institute of Standards and Technology
Building and Fire Research Laboratory
Gaithersburg, MD 20899-1070

J. Dreisbach, NRC Project Manager

NRC Job Code N6414

Office of Nuclear Regulatory Research

AVAILABILITY OF REFERENCE MATERIALS IN NRC PUBLICATIONS

NRC Reference Material

As of November 1999, you may electronically access NUREG-series publications and other NRC records at NRC's Public Electronic Reading Room at http://www.nrc.gov/reading-rm.html. Publicly released records include, to name a few, NUREG-series publications; *Federal Register* notices; applicant, licensee, and vendor documents and correspondence; NRC correspondence and internal memoranda; bulletins and information notices; inspection and investigative reports; licensee event reports; and Commission papers and their attachments.

NRC publications in the NUREG series, NRC regulations, and *Title 10, Energy*, in the Code of *Federal Regulations* may also be purchased from one of these two sources.
1. The Superintendent of Documents
   U.S. Government Printing Office
   Mail Stop SSOP
   Washington, DC 20402-0001
   Internet: bookstore.gpo.gov
   Telephone: 202-512-1800
   Fax: 202-512-2250
2. The National Technical Information Service
   Springfield, VA 22161-0002
   www.ntis.gov
   1-800-553-6847 or, locally, 703-605-6000

A single copy of each NRC draft report for comment is available free, to the extent of supply, upon written request as follows:
Address: Office of Administration, Reproduction and
         Distribution Services Section
         U.S. Nuclear Regulatory Commission
         Washington, DC 20555-0001
E-mail: DISTRIBUTION@nrc.gov
Facsimile: 301-415-2289

Some publications in the NUREG series that are posted at NRC's Web site address http://www.nrc.gov/reading-rm/doc-collections/nuregs are updated periodically and may differ from the last printed version. Although references to material found on a Web site bear the date the material was accessed, the material available on the date cited may subsequently be removed from the site.

Non-NRC Reference Material

Documents available from public and special technical libraries include all open literature items, such as books, journal articles, and transactions, *Federal Register* notices, Federal and State legislation, and congressional reports. Such documents as theses, dissertations, foreign reports and translations, and non-NRC conference proceedings may be purchased from their sponsoring organization.

Copies of industry codes and standards used in a substantive manner in the NRC regulatory process are maintained at—
   The NRC Technical Library
   Two White Flint North
   11545 Rockville Pike
   Rockville, MD 20852-2738

These standards are available in the library for reference use by the public. Codes and standards are usually copyrighted and may be purchased from the originating organization or, if they are American National Standards, from—
   American National Standards Institute
   11 West 42nd Street
   New York, NY 10036-8002
   www.ansi.org
   212-642-4900

Legally binding regulatory requirements are stated only in laws; NRC regulations; licenses, including technical specifications; or orders, not in NUREG-series publications. The views expressed in contractor-prepared publications in this series are not necessarily those of the NRC.

The NUREG series comprises (1) technical and administrative reports and books prepared by the staff (NUREG–XXXX) or agency contractors (NUREG/CR–XXXX), (2) proceedings of conferences (NUREG/CP–XXXX), (3) reports resulting from international agreements (NUREG/IA–XXXX), (4) brochures (NUREG/BR–XXXX), and (5) compilations of legal decisions and orders of the Commission and Atomic and Safety Licensing Boards and of Directors' decisions under Section 2.206 of NRC's regulations (NUREG–0750).

DISCLAIMER: This report was prepared as an account of work sponsored by an agency of the U.S. Government. Neither the U.S. Government nor any agency thereof, nor any employee, makes any warranty, expressed or implied, or assumes any legal liability or responsibility for any third party's use, or the results of such use, of any information, apparatus, product, or process disclosed in this publication, or represents that its use by such third party would not infringe privately owned rights.

ABSTRACT

This report describes a thermally-induced electrical failure (THIEF) model's ability to predict the behavior of power, instrument, and control cables during a fire. The model is intended to be incorporated as a subroutine for deterministic fire models, and it is of comparable accuracy and simplicity to the activation algorithms for various other fire protection devices (e.g., sprinklers, heat and smoke detectors). THIEF model predictions are compared to experimental measurements of instrumented cables in a variety of configurations, and the results indicate that the model is an appropriate analysis tool for nuclear power plant applications. This work was performed as part of the CAROLFIRE (Cable Response to Live Fire) program sponsored by the U.S. Nuclear Regulatory Commission. The experiments for CAROLFIRE were conducted at Sandia National Laboratories, Albuquerque, New Mexico. Details of the CAROLFIRE experimental program are contained in Volumes 1 and 2 of this three-volume series.

FOREWORD

The Browns Ferry fire in 1975 demonstrated that instrument, control and power cables are susceptible to fire damage. At Browns Ferry, over 1,600 cables were damaged by the fire and caused short circuits between energized conductors. These short circuits (i.e., "hot-shorts") caused certain systems to operate in an unexpected manner. Additionally, recent advances in the use of risk-informed methods indicate that hot-shorts, under certain circumstances, can pose a significant risk, and that plant risk analyses should account for those additional risks.

In order to better understand the issue of cable hot-shorts, the nuclear industry (Nuclear Energy Institute/Electric Power Research Institute) conducted a series of cable fire tests that were witnessed by the NRC staff in 2001. Based on the results of those tests, and data from previous tests available in the literature, the NRC facilitated a workshop on February 19, 2003. The workshop led the NRC to issue Regulatory Issue Summary (RIS) 2004-03, Revision 1, "Risk-Informed Approach for Post-Fire Safe Shutdown Circuit Inspections," December 29, 2004 (ADAMS Accession No. ML042440791), which describes the guidance NRC inspectors currently follow in deciding which causes of fire-induced hot-shorts are important to safety and should be considered during inspections. The RIS also describes "Bin 2" items, which are scenarios where the importance to safety of cable hot shots was unknown at the time of the workshop.

This report describes the CAROLFIRE (CAble Response tO Live FIRE) testing program. The primary objective of this program was to determine the safety importance of these Bin 2 items. A secondary objective of CAROLFIRE was to foster the development of cable thermal response and electrical failure fire modeling tools. To achieve these objectives, Sandia National Laboratories conducted a variety of fire experiments designed to examine the "Bin 2" items, and designed to capture cable thermal response and failure data. The cable thermal response data has been provided to the National Institute of Standards and Technology and the University of Maryland for use as the basis of development and initial validation of cable target response models.

The results presented in this report were from a series of both small- and intermediate-scale cable fire tests. The combined test matrices comprised 96 individual experiments of varying complexity. The tests involved a variety of common cable constructions and variations in test conditions like thermal exposure, raceway type, and bundling of similar and dissimilar cable types. The results provide the most extensive set of cable thermal response and failure data to date. This research provides valuable information and insights that may be used to evaluate the risk of fire-induced cable hot-shorts.

Christiana H. Lui, Director
Division of Risk Analysis
Office of Nuclear Regulatory Research
U.S. Nuclear Regulatory Commission

CONTENTS

Abstract iii
Foreword v
Contents vii
List of Figures ix
List of Tables xi
Executive Summary xiii
Acknowledgements xv
Abbreviation xvii
1 Background 1
   1.1 Introduction and Purpose 1
   1.2 A Brief History of Cable Modeling 1
2 Model Development 5
   2.1 Model Assumptions and Governing Equations 5
   2.2 Cable Properties 6
   2.3 Numerical Algorithm 9
      Finite Difference Scheme 9
      External Heat Flux 10
      Conduits 10
3 Bench-Scale "Penlight" Experiments 13
   3.1 Experimental Description 13
   3.2 Experimental Results 14
   3.3 Modeling Considerations 17
   3.4 Modeling Results 18
      Thermoset Cables 19
      Thermoplastic Cables 23
      Special Cables 27
   3.5 Summary of Penlight Analysis 28
   3.6 Why Does the THIEF Model Work? 32
4 Intermediate Scale Experiments 35
   4.1 Experimental Description 35
   4.2 Modeling Considerations 35
   4.3 Model Results 38
      Single Cables within Trays 38
      Cables in Conduits 40
      Random Fill Cable Trays 43
      Air Drops 45
      Six Cable Bundles 46
      Twelve Cable Bundles 51
   4.4 Summary of the Intermediate Scale Tests 53
5. Lessons Learned from CAROLFIRE 57
   5.1 Characterizing the thermal environment 57
   5.2 Specifying a "failure temperature" 57
   5.3 Defining the cable within the fire model 57
   5.4 Modeling cable burning 58

6	Conclusion	59
7	References	61

LIST OF FIGURES

Figure Title: Page

Figure 1. Curves representing the current guidance given in NUREG-1805 for estimating the time to cable electrical failure for a particular exposing temperature. Also shown are the results of the CAROLFIRE bench-scale (Penlight) experiments for single cables (Nowlen and Wyant 2007b). 3

Figure 2. Schematic diagram of Penlight apparatus, courtesy Sandia National Laboratory. Details can be found in Nowlen and Wyant (2007b). 13

Figure 3. Thermocouple locations for a 7/C cable. 14

Figure 4. The results of Penlight Test 2 where it is observed that the inner cable temperatures (solid and dashed red lines) increase significantly about 1 min prior to the first recorded electrical short. The dotted lines are results of the THIEF model. 15

Figure 5. (Left) Photograph of a conduit within the penlight apparatus, with end caps installed, courtesy Frank Wyant, Sandia National Laboratories. (Right) Predicted (dotted line) and measured (solid line) temperatures of the conduit when exposed to the shroud temperature profile. This example is part of penlight experiment number 7. 17

Figure 6. Key to graphs showing results of penlight tests. Note that the term "First Short" indicates the first observed electrical failure, regardless of the specific type. In the experiments, subsequent shorts were observed, but are not relevant here. 18

Figure 7. Results for Cable #14 (XLPE/CSPE, 3 conductor). The upper-most black curves depict the measured (solid) and specified (dotted) shroud temperature. The solid and dashed red lines are the measured temperatures inside the cable jacket, at opposite sides. The dotted red curve is the predicted inner cable temperature. The vertical dashed line indicates the first observed electrical short in the experiment. For Test 7, the solid blue and dotted blue curves represent the measured and predicted conduit temperatures, respectively. 20

Figure 8. Results for Cable #10 (XLPE/CSPE, 7 conductor). The color scheme is the same as in Figure 7. 21

Figure 9. Results for various thermoset cables. The upper-most black curves depict the measured (solid) and specified (dotted) shroud temperature. The solid and dashed red lines are the measured temperatures inside the cable jacket, at opposite sides. The dotted red curve is the predicted inner cable temperature. The vertical dashed line indicates the first observed electrical short in the experiment. 22

Figure 10. Results for Cable #5 (PVC/PVC, 3 conductor). The upper-most black curves depict the measured (solid) and specified (dotted) shroud temperature. The solid and dashed red lines are the measured temperatures inside the cable jacket, at opposite sides. The dotted red curve is the predicted inner cable temperature. The vertical dashed line indicates the first observed electrical short in the experiment. For Test 8, the solid blue and dotted blue curves represent the measured and predicted conduit temperatures, respectively. 24

Figure 11. Results for Cable #15 (PE/PVC, 7 conductor). The color scheme is the same as in Figure 10. 25

Figure 12. Results for 2, 7, and 12 conductor PVC/PVC (thermoplastic) cables. The upper-most black curves depict the measured (solid) and specified (dotted) shroud temperature. The solid and dashed red lines are the measured temperatures inside the cable jacket, at opposite sides. The dotted red curve is the predicted inner cable temperature. The vertical dashed line indicates the first observed electrical short in the experiment. 26

Figure 13. Results for Cable #9 (Silicone Rubber/Aramid Braid, 7 conductor) and Cable #11 (VITA-LINK, 7 conductor). The upper-most black curves depict the measured (solid) and model-specified (dotted) shroud temperature. The solid red line is the measured temperature inside the cable jacket. The dotted red line is the predicted inner cable temperature. Ignition and burning are not included in the THIEF model, and electrical shorting did not occur in these two experiments. Note that neither cable failed electrically until the introduction of a water spray. See Nowlen and Wyant (2007b) for details. 27

Figure 14. Comparison of THIEF model and experiment for the Penlight series. Point A and B refer to the thermocouple measurements made on opposite sides of the cable, just under the jacket. The "Threshold Temperature" is 400 °C (752 °F) for thermoset cables and 200 °C (392 °F) for thermoplastics. The dashed lines represent the average (-3 %) and standard deviation (20 %) of the data. 29

Figure 15. Sensitivity of the THIEF model predictions to various cable properties. The Relative Difference refers to the predicted vs measured times to reach the threshold temperature. The horizontal dashed line is the average Relative Difference for all the Penlight comparisons. 31

Figure 16. Results of an alternative thermal model (short dashes) and the THIEF model (dotted line), compared to measurements (solid and long dashed lines). 32

Figure 17. Side view of the Intermediate Scale Test rig. Courtesy Sandia National Laboratories. 35

Figure 18. Key to Intermediate Scale graphs. 36

Figure 19. Example of an experiment where the temperature measurement (TC-1) may not coincide with the location of the first electrical short. 37

Figure 20. Summary of results for single, isolated cables in the Intermediate scale tests. Black indicates the exposing temperature; red the cable. Solid lines for experiment, dotted for model. 39

Figure 21. Summary of results for the cables in the conduit at Location E. Black indicates the exposing temperature; red the cable. Solid lines for experiment, dotted for model. 41

Figure 22. Summary of results for cables in conduits at Location D and G. Black indicates the exposing temperature; red the cable. Solid lines for experiment, dotted for model. 42

Figure 23. Cable tray configuration for Intermediate Test 1. 43

Figure 24. Cable tray configuration for Intermediate Tests 13 and 14. 43

Figure 25. Summary of results for trays filled with random cable. Black indicates the exposing temperature; red the cable. Solid lines for experiment, dotted for model. The measured temperature at Location A in Test 13 (middle left graph) is inconsistent with that at Location C (middle right graph). 44

Figure 26. Summary of results for air drop cables in the Intermediate scale tests. Black indicates the exposing temperature; red the cable. Solid lines for experiment, dotted for model. 45

Figure 27. Summary of results for 6 cable bundles, Location A. Black indicates the exposing temperature; red the cable. Solid lines for experiment, dotted for model. 47

Figure 28. Summary of results for 6 cable bundles, Locations A and C. Black indicates the exposing temperature; red the cable. Solid lines for experiment, dotted for model. 48

Figure 29. Summary of results for 6 cable bundles, Locations C and F. Black indicates the

exposing temperature; red the cable. Solid lines for experiment, dotted for model. 49
Figure 30. Summary of results for the 6 cable bundles, Location G. Black indicates the exposing temperature; red the cable. Solid lines for experiment, dotted for model. 50
Figure 31. Summary of results for the 12 cable bundles. Black indicates the exposing temperature; red the cable. Solid lines for experiment, dotted for model. 52
Figure 32. Summary of the Intermediate Scale Test predictions. The dashed lines indicate the average (-15 %) and standard deviation (33 %) of the data. Also note that the description "Inside Fire Plume" refers to locations A and C, which were within the flaming region of the fire. 54

LIST OF TABLES

Table Title: Page

Table 1. Physical properties of the cables that are relevant to the modeling study. Additional information is included in Nowlen and Wyant (2007a). Note that the "Cable Number" is used only to identify the cables used in the study 8
Table 2. Summary of the Sandia Penlight experiments that involved only single cables. Measurement uncertainties are reported in Nowlen and Wyant (2007b). Note that the thermoplastic cable (TEF/TEF) in Test 22 was tested like a thermoset. 16
Table 3. Summary of the THIEF Model Predictions of the Penlight Experiments. 30
Table 4. Results of the Intermediate Scale THIEF Model Predictions. See Nowlen and Wyant (2007b) for details about each configuration. 55

EXECUTIVE SUMMARY

This report documents the numerical modeling results from the Cable Response to Live Fire (CAROLFIRE) project. CAROLFIRE is a U.S. Nuclear Regulatory Commission (US NRC) Office of Nuclear Regulatory Research (RES) initiated effort to study the fire-induced thermal response and functional behavior of electrical cables. The project is a collaborative effort that includes the NRC Office of Nuclear Reactor Regulation (NRR) as peer reviewers, Sandia National Laboratories (SNL) as the primary testing laboratory, and both the University of Maryland and the National Institute of Standards and Technology (NIST) as general collaborative partners to develop a better predictive model for cable thermal response in deterministic fire models.

The primary project objective of CAROLFIRE is to characterize the various modes of electrical failure (*e.g.* hot shorts, shorts to ground) within bundles of power, control and instrument cables. A secondary objective of the project is to develop a simple model to predict Thermally-Induced Electrical Failure (THIEF) when a given interior region of the cable reaches an empirically determined threshold temperature. The measurements used for these purposes are described in Volume II of the CAROLFIRE test report (Nowlen and Wyant 2007b).

The THIEF model for cables has been shown to work effectively in realistic fire environments. The THIEF model is essentially nothing more than the numerical solution of the one-dimensional heat conduction equation within a homogenous cylinder with fixed, temperature-independent properties. The model was used to predict the inner cable temperature of 100 instrumented cables from the CAROLFIRE Penlight (35 single cable experiments; 66 point to point comparisons) and Intermediate Scale Test Series (14 experiments; 65 point to point comparisons). Because the Penlight experiments tested single cables that were heated uniformly on all sides, the one-dimensional THIEF model accurately predicted the times for the temperature inside the cable jacket to reach "threshold" values that are typically observed when the cable fails electrically. For 66 measurements, the model under-predicted the time to reach threshold temperature by 3 %, on average. In the Intermediate Scale experiments, where the cable configurations were more typical of actual installations, the model under-predicted the times to reach threshold temperature by 15 %, on average. This latter result is realistically conservative – the THIEF model does not account for the shielding effects of cable bundles, and thus over-predicts cable temperatures and under-predicts "failure" times.

ACKNOWLEDGEMENTS

The work described in this report was supported by the Office of Nuclear Regulatory Research (RES) of the US Nuclear Regulatory Commission (USNRC). The CAROLFIRE program was directed by H.W (Roy) Woods and Mark Henry Salley, with assistance from Jason Dreisbach and Felix Gonzalez. The experiments described in this report were conducted by Steve Nowlen and Frank Wyant of Sandia National Laboratories.

Special thanks to Petra Andersson and Patrick Van Hees at SP, Sweden, for their development of the thermally-induced electrical failure (THIEF) model that is described in this report.

ABBREVIATION

AWG	American Wire Gauge
CAROLFIRE	Cable Response to Live Fire
CPE	Chlorinated Polyethylene
CSPE	Chloro-Sulfanated Polyethylene
EPR	Ethylene-Propylene Rubber
EPRI	Electric Power Research Institute
NFPA	National Fire Protection Association
NIST	National Institute of Standards and Technology
NRC	Nuclear Regulatory Commission
NRR	NRC Office of Nuclear Reactor Regulation
PE	Polyethylene
PVC	Poly-vinyl Chloride
RES	NRC Office of Nuclear Regulatory Research
SNL	Sandia National Laboratories
SR	Silicone Rubber
TC	Thermocouple
THIEF	Thermally-Induced Electrical Failure
TP	Thermoplastic
TS	Thermoset
XLPE	Cross-Linked Polyethylene
XLPO	Cross-Linked Polyolefin

1 BACKGROUND

1.1 Introduction and Purpose

The <u>Ca</u>ble <u>R</u>esponse to <u>L</u>ive <u>Fi</u>re (CAROLFIRE) project is a U.S. Nuclear Regulatory Commission (US NRC) Office of Nuclear Regulatory Research (RES) initiated effort to study the fire-induced thermal response and functional behavior of electrical cables. The project is a collaborative effort that includes the NRC Office of Nuclear Reactor Regulation (NRR) as peer reviewers, Sandia National Laboratories (SNL) as the primary testing laboratory, and both the University of Maryland and the National Institute of Standards and Technology (NIST) as general collaborative partners to develop a better predictive model for cable thermal response in deterministic fire models.

The primary project objective of CAROLFIRE is to characterize the various modes of electrical failure (*e.g.* hot shorts, shorts to ground) within bundles of power, control and instrument cables. Details can be found in Volume 1 of the CAROLFIRE test report (Nowlen and Wyant 2007a). A secondary objective of the project is to develop a simple model to predict <u>Th</u>ermally-<u>I</u>nduced <u>E</u>lectrical <u>F</u>ailure (THIEF) when a given interior region of the cable reaches an empirically determined threshold temperature. The measurements used for these purposes are described in Volume II of the CAROLFIRE test report (Nowlen and Wyant 2007b).

The description and validation of the THIEF model are reported here. Current nuclear power plant probabilistic risk assessment (PRA) methods employ simple linear regression techniques to predict cable performance in a fire. These methods take into account the general composition of the cable, but not other information, like its mass or diameter. The THIEF model described in this report uses the general cable construction and bulk properties, but does not require more detailed thermo-physical properties. For example, the mass per unit length and diameter are needed, but the thermal conductivity, specific heat, and emissivity are assumed, based on the current generation of cables in existing plants. This latter detailed information is not always readily available for the wide variety of often proprietary cable materials, and bench-scale experiments to measure the properties can be expensive and difficult to perform for all existing and future cable materials.

1.2 A Brief History of Cable Modeling

The thermal decomposition and electrical failure[1] of multi-conductor cables in a fire have been of interest to the nuclear power industry dating back to the Browns Ferry fire of 1975 (US NRC 1975). However, the development of a predictive model of cable failure has been elusive for a number of reasons. First, cables are a fairly complex combination of insulating plastics, metal conductors, protective armors, and a variety of filler materials. The availability of comprehensive thermo-physical properties of these materials is limited. Even when the material

---

[1] Throughout this report, the term "electrical failure" applies to any situation where a cable no longer functions as designed due to the heating by a fire. Most often, these failures come in the form of hot shorts or shorts to ground that can result in spurious actuations of system and components. See Nowlen and Wyant (2007a) for a more detailed discussion.

properties for a particular cable are available, it is still a challenge to calculate the heat penetration through a bundle of the cables lying in a tray or run through a conduit.

Rather than try to develop detailed models, engineers have looked for a practical correspondence between electrical failure and the compartment temperature in a fire. A simple approach is to develop an empirical relationship between the time to electrical failure and the "exposing" temperature; that is, the temperature of the hot gases in the vicinity of the cable. NUREG-1805 (Iqbal and Salley 2004), a set of engineering calculation methods specifically designed for nuclear power plant applications, suggests that the time to electrical failure is inversely proportional to the exposing temperature. For the two major classes of cables, thermosets and thermoplastics,[2] it provides an estimated failure time (in seconds) for a given exposing temperature (in °C):

$$\frac{1}{t} = 3.343 \times 10^{-5} T - 0.01044 \qquad \text{for thermoset cables} \qquad (1.1)$$

$$\frac{1}{t} = 3.488 \times 10^{-5} T - 0.007467 \qquad \text{for thermoplastic cables} \qquad (1.2)$$

These two curves are shown in Figure 1, along with some results from the CAROLFIRE experimental program (Nowlen and Wyant 2007b). While Eqs. (1.1) and (1.2) are useful screening methods, they are somewhat limited in application. First, they are based on constant temperature exposures, which is unrealistic in fire. For example, the CAROLFIRE results shown in Figure 1 are only from the small-scale experiments in which individual cables were exposed to a uniform temperature. The larger-scale test results from CAROLFIRE cannot be characterized in terms of a single exposing temperature or heat flux. Second, Eqs. (1.1) and (1.2) do not account for different cable installations or configurations. For example, suppose the cable is routed through a conduit, or has a protective armor jacket. What if the cable is considerably different in size and composition to those that were tested? The formulae only distinguish between a thermoset and thermoplastic cable, based on the fact that the latter have been shown to fail at lower temperatures than the former. The formulae do not take into account size, mass, protective barriers, or site-specific conditions. For several of the CAROLFIRE test configurations, neither Eqs. (1.1) nor (1.2) could be applied directly.

Because of these limitations, a more flexible predictive model must have some consideration for the thermal mass of the cable, and it must infer electrical failure from the attainment of a given "failure" temperature somewhere within the cable. Over the past 30 years, a number of studies on electrical cable performance in fires have suggested various "failure" temperatures for different classes of cables. A review of these studies is included in NUREG-1805 (Iqbal and Salley 2004). The intent of this report is to demonstrate that a simple heat conduction calculation, along with an empirically-based "failure" temperature, is sufficient to predict cable

---

[2] Plastics can be classified into two major categories: thermoplastics and thermosets. In general, thermoplastics can be heated, melted and cooled to solid form. Thermosets will reach decomposition temperature before melting temperature and will degrade irreversibly if exposed to high temperatures.

failure times to an accuracy that is consistent with that of current generation fire models (US NRC and EPRI 2007). The calculation is described in the next section.

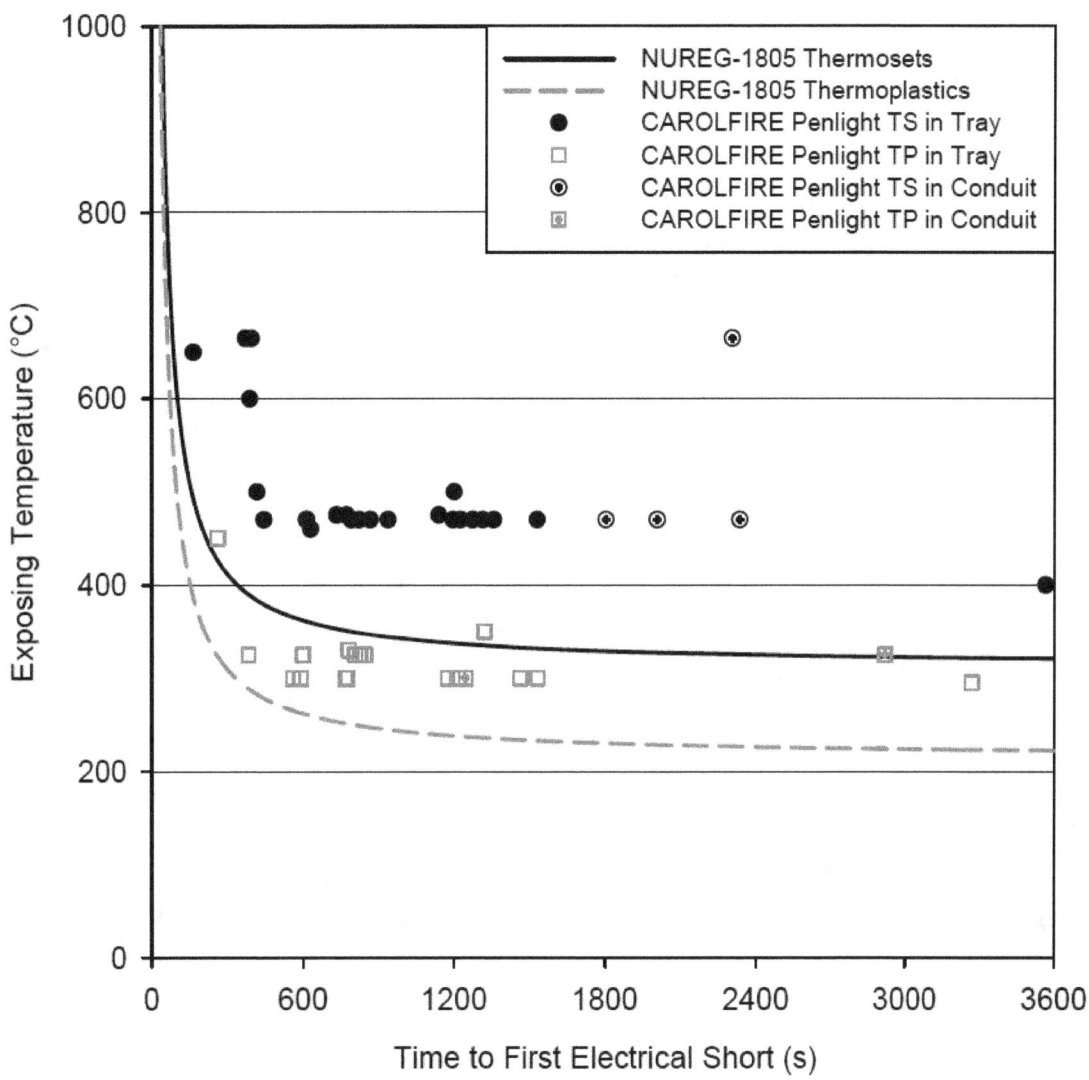

Figure 1. Curves representing the current guidance given in NUREG-1805 for estimating the time to cable electrical failure for a particular exposing temperature. Also shown are the results of the CAROLFIRE bench-scale (Penlight) experiments for single cables (Nowlen and Wyant 2007b).

2 MODEL DEVELOPMENT

Petra Andersson and Patrick Van Hees of the Swedish National Testing and Research Institute (SP) have proposed that a cable's <u>th</u>ermally-<u>i</u>nduced <u>e</u>lectrical <u>f</u>ailure (THIEF) can be predicted via a simple one-dimensional heat transfer calculation, under the assumption that the cable can be treated as a homogenous cylinder (Andersson and Van Hees 2005). Their results for PVC cables were encouraging and suggested that the simplification of the analysis is reasonable and that it should extend to other types of cables. In the section, the model is described.

2.1 Model Assumptions and Governing Equations

The assumptions underlying the THIEF model are as follows:

1. The heat penetration into a cable of circular cross section is largely in the radial direction. This greatly simplifies the analysis, and it is also conservative because it is assumed that the cable is completely surrounded by the heat source.
2. The cable is homogenous in composition. In reality, a cable is constructed of several different types of polymeric materials, cellulosic fillers, and a conducting metal, most often copper.
3. The thermal properties – conductivity, specific heat, and density – of the assumed homogenous cable are independent of temperature. In reality, both the thermal conductivity and specific heat of polymers are temperature-dependent, but this information is very difficult to obtain from manufacturers. More discussion of this assumption is found below.
4. It is assumed that no decomposition reactions occur within the cable during its heating, and ignition and burning are not considered in the model. In fact, thermoplastic cables melt, thermosets form a char layer, and both off-gas volatiles up to and beyond the point of electrical failure.
5. Electrical failure occurs when the temperature just inside the cable jacket reaches an experimentally determined value.

Given these assumptions, the governing equation for the cable temperature, $T(r,t)$, is given by:

$$\rho c \frac{\partial T}{\partial t} = \frac{1}{r} \frac{\partial}{\partial r} k \, r \frac{\partial T}{\partial r} \qquad (2.1)$$

where ρ, c, and k are the effective density, specific heat, and conductivity of the <u>s</u>olid, all assumed constant. The boundary condition at the exterior boundary, $r = R$, is given by:

$$k \frac{\partial T}{\partial r}(R,t) = \dot{q}'' \qquad (2.2)$$

where $\dot{q}''$ is the assumed axially-symmetric heat flux to the exterior surface of the cable. The heat flux is provided by the fire model or fire analysis that is being used to assess the overall thermal environment of the compartment where the cable is located. In most realistic fire scenarios, the heat flux to the cable is not axially-symmetric. For the purpose of modeling cable failure, it is recommended that the maximum value of the heat flux be used.

Obviously, there are considerable assumptions inherent in the Andersson and Van Hees THIEF model, but their results for various polyvinyl chloride (PVC) cables suggest that it may be sufficient for engineering analyses of a wider variety of cables. In this report, the model is applied to fifteen different cable samples that have been exposed to a variety of thermal exposures. The only difference in the application of the model here is that the 1-D heat transfer equation (2.1) is solved numerically rather than analytically. The analytical solution derived by the SP researchers, while perfectly correct, is fairly complicated and a simple numerical solution is easier to implement in a large-scale fire model. Indeed, most fire models already employ a 1-D heat transfer algorithm to compute heat losses to walls. The accuracy of either the analytical solution or the numerical solution is not of concern, given the much greater uncertainty in the material properties of the plastic and the underlying assumption of homogeneity. Moreover, the numerical solution is less restricted, which is important if it is found that a particular type of cable cannot be described as a homogenous cylinder.

The THIEF model can only predict the temperature profile within the cable as a function of time, given a time-dependent exposing temperature or heat flux. The model does not predict at what temperature the cable fails electrically. This information is gathered from experiment. The CAROLFIRE experimental program included bench-scale, single cable experiments in which temperature measurements were made on the surface of, and at various points within, cables subjected to a uniform heat flux. These experiments provided the link between internal cable temperature and electrical failure. The model can only predict the interior temperature and infer electrical failure when a given "failure" temperature is reached. It is presumed that the temperature of the centermost point in the cable is not necessarily the indicator of electrical failure. This analysis method uses the temperature just inside the cable jacket rather than the centermost temperature, as that is where electrical shorts in a multi-conductor cable are most likely to occur first.

2.2 Cable Properties

Fifteen types of cable construction were tested in the CAROLFIRE project. A detailed description of each can be found in Nowlen and Wyant (2007a). Each cable is typically composed of an outer jacket, insulated conductors, and, for certain types, a light weight filler material. Various polymers are used for the jacket and insulation, typically classified as either thermoset or thermoplastic. The THIEF model does not distinguish between thermosets and thermoplastics, but the behavior of each at elevated temperatures is distinctly different, and consequently the experimental test parameters depended on the cable type.

Table 1 provides information about the cables that is relevant to the THIEF model. The material names have been abbreviated as follows:

Aramid Braid ®	A synthetic fiber manufactured by DuPont Chemical (Thermoset)
CPE	Chlorinated Polyethylene (Thermoset)

CSPE	Chlorosulfonated Polyethylene (Thermoset)
EPR	Ethylene Propylene Rubber (Thermoset)
PE	Polyethylene (Thermoplastic)
PVC	Polyvinyl Chloride (Thermoplastic)
SR	Silicone-Rubber (Thermoset)
Tefzel ®	A fluoropolymer resin manufactured by DuPont (Thermoplastic)
Vitalink ®	A silicone rubber manufactured by Rockbestos-Surprenant (Thermoset)
XLPE	Cross-Linked Polyethylene (Thermoset)
XLPO	Cross-Linked Polyolefin (Thermoset)

The only information required by the THIEF model for a particular cable is its overall diameter, its mass per unit length, its outer jacket thickness, and an experimentally determined "failure" temperature. The first two pieces of information are needed to describe the geometry and bulk thermal inertia of the cable; the jacket thickness is needed because it is assumed that cables fail electrically when the temperature of the insulation material surrounding the first layer of conductors just inside the jacket of a multi-conductor cable reaches a particular value which is determined experimentally (the last piece of information required as input). The cable diameter, mass per unit length and jacket thickness are all easily obtained either from the manufacturer or by direct measurement. The insulation thickness itself is not used by the model, but it is included in Table 1 because it might provide some insight into the failure mechanism. Generally speaking, the insulation is relatively thin, and its thermal penetration time is relatively short, compared to the jacket. The copper volume and mass fractions are included only to demonstrate the varied make up of the different cables. It will be shown below that the results of the THIEF model are insensitive to the relative amounts of copper and plastic, at least for the 15 cable types tested.

One of the challenges in developing a more detailed model of the cable is the difficulty in obtaining material properties. The insulation and jacket materials are often complex polymers that undergo a number of reactions as they heat. Given the complexity of these processes and the expense of obtaining various thermo-physical properties, the THIEF model employs a single value for the specific heat and the thermal conductivity, 1.5 kJ/kg/K and 0.2 W/m/K, respectively, for both thermoset and thermoplastic cables. The bulk density of the cable, ρ, can be calculated by dividing the inputted mass per unit length by the cross sectional area. The emissivity of the cable jacket is assumed to be 0.95. These values are typical of several types of commonly used cable jacket and insulation materials, as reported by Hamins *et al.* (2006). Of course, each type of polymer is different; the properties are temperature-dependent, other decomposition reactions occur, *etc.* The calculations presented in this report could easily be repeated using other values, but it is hardly worthwhile because the predicted "failure" times are, to a first approximation, linearly proportional to the jacket thickness, the specific heat, and the density, and inversely proportional to the thermal conductivity.

Table 1. Physical properties of the cables that are relevant to the modeling study. Additional information is included in Nowlen and Wyant (2007a). Note that the "Cable Number" is used only to identify the cables used in the study.

Cable Number	Cable Description (Insulation/Jacket/ No. Conductors)	Cable Classification	Insulation Thickness (mm)	Jacket Thickness (mm)	Outer Diameter (mm)	Mass per Length (kg/m)	Bulk Density (kg/m^3)	Copper Volume Fraction	Copper Mass Fraction
1	PVC/PVC, 7/C	Thermoplastic	0.5	1.1	12.4	0.324	2680	0.24	0.80
2	EPR/CPE, 7/C	Thermoset	0.8	1.5	15.1	0.400	2232	0.16	0.65
3	XLPE/PVC, 7/C	Mixed	0.8	1.5	15.1	0.388	2166	0.16	0.67
4	PVC/PVC, 2/C	Thermoplastic	0.7	1.0	7.0	0.076	1668	0.07	0.40
5	PVC/PVC, 3/C	Thermoplastic	0.9	1.5	15.2	0.459	2532	0.18	0.63
6	PVC/PVC, 12/C	Thermoplastic	0.5	1.1	11.3	0.195	1948	0.13	0.59
7	XLPE/CSPE, 2/C	Thermoset	0.6	1.1	7.9	0.097	1993	0.07	0.31
8	XLPO/XLPO, 7/C	Thermoset	0.5	0.9	12.2	0.321	2743	0.25	0.81
9	SR/Aramid Braid, 7/C	Thermoset	1.3	1.0	14.5	0.358	2168	0.18	0.73
10	XLPE/CSPE, 7/C	Thermoset	0.8	1.5	15.0	0.410	2321	0.16	0.63
11	VITA-LINK, 7/C	Thermoset	1.5	2.0	19.0	0.500	1656	0.06	0.34
12	TEF/TEF, 7/C	Thermoplastic	0.4	0.5	10.2	0.292	3578	0.36	0.89
13	XLPE/CSPE, 12/C	Thermoset	0.6	1.1	12.7	0.231	1825	0.10	0.50
14	XLPE/CSPE, 3/C	Thermoset	1.1	1.5	16.3	0.529	2538	0.16	0.55
15	PE/PVC, 12/C	Thermoplastic	0.8	1.1	15.0	0.380	2152	0.16	0.68

2.3 Numerical Algorithm

This section provides details of the numerical solution of the heat conduction equation described in Section 2.1. It can be incorporated into any fire model that predicts the thermal environment surrounding the cable(s). This can be as simple as a gas temperature predicted by an empirical correlation, or as detailed as a spatially-resolved flow field in a computational fluid dynamics model. Whatever model is chosen, it must produce an estimate, as a function of time, of the heat flux to the cable surface, even if the cable itself is not explicitly included in the fire model.

Finite Difference Scheme

To solve Eq. (2.1) numerically, first divide the radius, R, of the cable into N uniformly spaced increments of length, $\delta r = R/N$. An appropriate value for δr is about 0.1 mm for cables similar to those tested in CAROLFIRE. Next, define a time step that is related to the spatial increment. This is known as the *time step constraint*, which is necessary for accuracy, and sometimes numerical stability:

$$\delta t = \frac{c \rho \, \delta r^2}{2k} \tag{2.3}$$

The temperature of the i-th radial increment (or *cell*) at the n-th time step ($t^n = n \, \delta t$) is denoted, T_i^n. The value of the radius at the forward edge of the i-th cell is denoted, $r_i = i \, \delta r$. Thus, $r_0 = 0$ and $r_N = R$.

A finite difference approximation to Eq. (2.1), *second order accurate in space and time*[3], is given by:

$$\rho c \frac{T_i^{n+1} - T_i^n}{\delta t} = \frac{2k}{(r_{i+1} + r_i)} \frac{1}{2\delta r} \left[r_i \frac{T_{i+1}^n - T_i^n}{\delta r} - r_{i-1} \frac{T_i^n - T_{i-1}^n}{\delta r} + r_i \frac{T_{i+1}^{n+1} - T_i^{n+1}}{\delta r} - r_{i-1} \frac{T_i^{n+1} - T_{i-1}^{n+1}}{\delta r} \right] \tag{2.4}$$

where ρ, c, and k are the effective density, specific heat, and conductivity of the solid, all assumed constant and given in the previous section. The boundary condition (Eq. (2.2)) is written in finite difference form as:

$$k \frac{T_{N+1}^n - T_N^n}{\delta r} = \dot{q}''(t^n) \tag{2.5}$$

Note that $T_{s,N+1}^n$ is a fictitious value of the temperature half a grid increment away from the exterior surface. It serves only to define the temperature gradient at the cable surface.

---

[3] The *order of accuracy* of the scheme has to do with the finite difference form of the partial derivatives in the original equation. The scheme shown here is known as the Crank-Nichsolson Method, and it was used to produce the results shown in this report. A simpler implementation, which is first order accurate in time instead of second, can be derived simply by replacing the superscripts $n+1$ by n on the right side of the equation. This has the effect of making the scheme *explicit*, meaning that the solution can be advanced in time by just moving all the terms defined at the n-th time step to the right hand side of the equation and solving for the temperature at the next ($n+1$) time step directly. The Crank-Nicholson Method requires the solution of a tri-diagonal system of linear equations.

External Heat Flux

The *net*[4] heat flux at the surface is determined from the exposing gas temperature surrounding the cable at the *n*-th time step, $T_g(t^n)$:

$$\dot{q}''(t^n) = \varepsilon \sigma \left(T_g(t^n)^4 - (T_s^n)^4 \right) + h \left(T_g(t^n) - T_s^n \right) \tag{2.6}$$

Here, ε is the emissivity of the cable surface (assumed to be 0.95 by THIEF), σ is the Stefan-Boltzmann constant (5.67×10^{-8} W/m²/K⁴), h is the convective heat transfer coefficient (assumed to be 10 W/m²/K, typical of free convection (Incropera and DeWitt 1990)), and $T_g(t^n)$ is the *effective*[5] gas temperature at the *n*-th time step, which is calculated by the fire model in which THIEF is embedded. The surface temperature of the cable, T_s^n, can be taken as T_N^n, the temperature one half of a cell width inside the cable surface. Typically, the cell width is chosen to be about 0.1 mm, a very small length, and this minor approximation has very little influence on the solution.

The fire model, in which THIEF is embedded as a *target sub-model*, may or may not produce the heat flux directly. At the very least, the fire model predicts an upper layer temperature, which for this application can be taken as the *effective* gas temperature, T_g, to which the cable is exposed. Depending on the application, the cable(s) might be exposed directly to the fire, in which case the required *net* radiative heat flux to the cable is taken as the incident heat flux determined by the fire model (point source method, for example) minus the radiative loss of the cable:

$$\dot{q}''(t^n) = \varepsilon \dot{q}''_{inc} - \varepsilon \sigma (T_N^n)^4 \tag{2.7}$$

Further discussion of the heat flux is included in Section 5, "Lessons Learned from CAROLFIRE," in which various issues related to the implementation of THIEF in a fire are discussed.

Conduits

A slight complication of the solution methodology described above is in situations where the cable is surrounded by a protective layer like a conduit, armor jacket, or tray covering. In CAROLFIRE, only conduits were considered in the modeling, but other protective measures can be handled in similar fashion, assuming test data is available to validate the various physical assumptions.

A conduit forms a thermal barrier between the hot gases of a fire and the cable itself. A simple way to incorporate its effect into the THIEF model is to replace the "exposing" gas temperature, T_g, in Eq. (2.6) by the conduit's temperature, T_c. In other words, the cable no longer "sees" the hot gases from the fire, but rather the interior surface of the conduit. A steel conduit may be

---
[4] The *net* flux implies the incident minus the re-radiated radiation heat flux.
[5] It is assumed that in most applications of the THIEF model, the cables are to be surrounded by hot, smoke-filled gases from a fire, and that these gases are optically-thick, or "black."

assumed thermally-thin; that is, its conductivity is so large that for all practical purposes it can be assumed that its exterior and interior surface temperatures are equal. Its temperature increases due to the heat flux from the hot gases at its exterior surface:

$$\dot{q}''_{ext}(t^n) = \varepsilon_c \sigma \left(T_g(t^n)^4 - (T_c^n)^4\right) + h\left(T_g(t^n) - T_c^n\right) \tag{2.8}$$

Here, T_c^n is the conduit temperature at the *n*-th time step and ε_c is the emissivity of its surface, taken as 0.85, typical of non-polished steel (Weast 1982). Heat is transferred from the interior surface of the conduit to the cable surface via:

$$\dot{q}''_{int}(t^n) = F\sigma\left((T_c^n)^4 - (T_N^n)^4\right) + h\left(T_c^n - T_N^n\right) \quad ; \quad F = \left(\left(\frac{R_c}{R}\right)\frac{1}{\varepsilon} + \frac{1-\varepsilon_c}{\varepsilon_c}\right)^{-1} \tag{2.9}$$

Here, R_c is the inner radius of the conduit. The view factor, F, is based on the assumption that the conduit and cable are concentric cylinders (Incropera and DeWitt 1990). The temperature of the conduit is raised by the net heat fluxes to its exterior and from its interior surfaces:

$$T_c^{n+1} = T_c^n + \delta t \frac{\dot{q}''_{ext} - \dot{q}''_{int}}{\rho_c c_c \delta_c} \tag{2.10}$$

Here, the subscript, *c*, stands for conduit, and the density and specific heat are that of steel (7850 kg/m³ and 0.46 kJ/kg/K). The thickness of the conduit, δ_c, used in the CAROLFIRE experiments was 4.9 mm (0.19 in).

3 BENCH-SCALE "PENLIGHT" EXPERIMENTS

The CAROLFIRE test program consisted of small-scale and intermediate-scale experiments, all conducted at Sandia National Laboratories. The experiments are described briefly below. Details can be found in Nowlen and Wyant (2007b). The information included in this report is that which is relevant to explain how the THIEF model of cable failure outlined above was applied for all 15 different cable construction samples for the various test configurations. This chapter describes the small-scale experiments in what is referred to as the "penlight" apparatus.

3.1 Experimental Description

The penlight is a cylinder formed by heating elements, 0.60 m (2 ft) long and 0.45 m (1.5 ft) in diameter, usually oriented horizontally, as shown in Figure 2. In the experiments, the temperature of the cylindrical "shroud" was controlled according to a specified function of time, while pairs of cables were monitored, one for thermal and the other for electrical response.

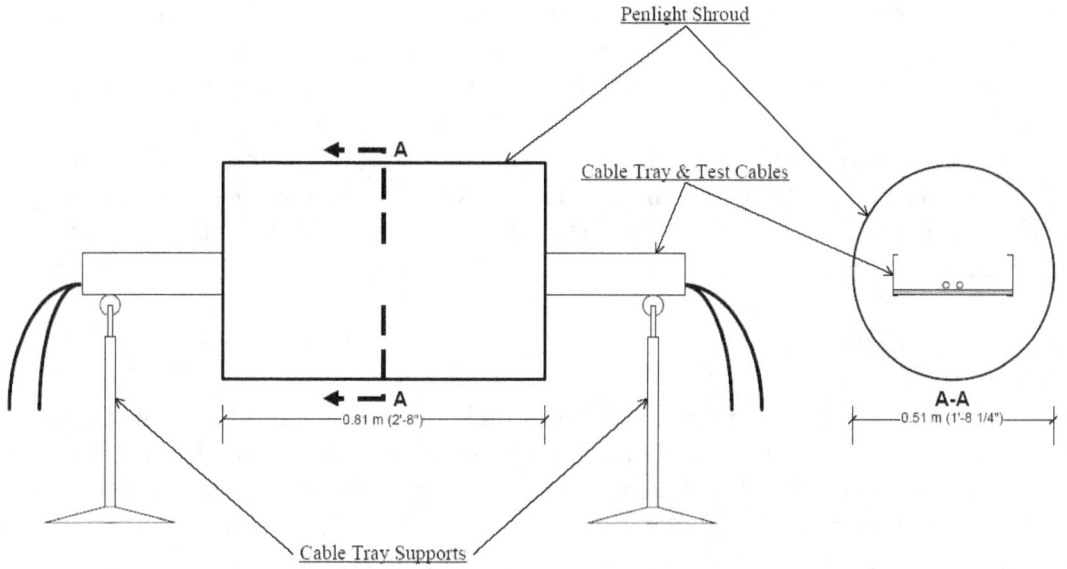

Figure 2. Schematic diagram of Penlight apparatus, courtesy Sandia National Laboratory. Details can be found in Nowlen and Wyant (2007b).

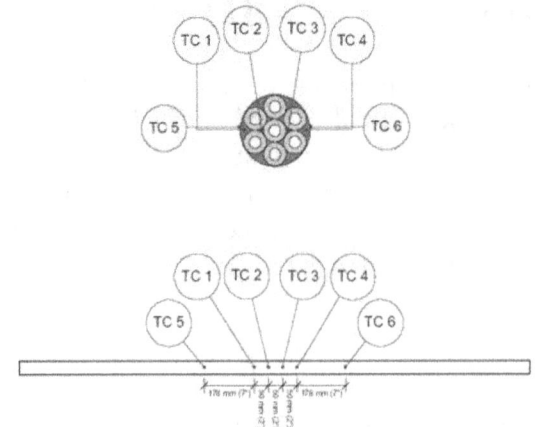

Figure 3. Thermocouple locations for a 7/C cable.

The figure at left shows the thermocouple locations for a seven conductor (7/C) cable. The thermocouples were affixed to the cable surface and inserted inside the cable, just under the jacket, at various locations, depending on the number of conductors. For comparison to the THIEF model, only the two measurements just inside the jacket (TC 1 and TC 4) were used. These measurements were made very near the center of the penlight apparatus, where the heat flux was expected to be greatest.

There were 35 Penlight experiments in which single cables were monitored for thermal and electrical response, designated in Nowlen and Wyant (2007b) as PT-1 through PT-31 and PT-62 through PT-65. There were three ways in which the cables were supported within the penlight apparatus: on a tray, in a conduit, or merely suspended midway through the penlight tunnel, commonly referred to as an "air drop". Other penlight experiments were conducted with multiple cables bundled together and monitored electrically. The primary purpose of these experiments was to address the various modes of electrical cable failure; they were not designed to assess the thermal response of a set of bundled cables. Consequently, the issue of predicting the temperatures of bundled cables is addressed in the chapter describing the Intermediate Scale Test Series.

3.2 Experimental Results

The results of 35 penlight experiments involving single (non-bundled) cables are summarized in Table 2. From the measurements of the temperatures just below the jacket, it is fairly evident that the tested thermoplastic cables failed electrically when their inner (under the jacket) temperatures reached somewhere between 200 °C and 250 °C (392 °F and 482 °F). For thermosets, the range was about 400 °C to 450 °C (752 °F to 842 °F). It is not possible to be more precise for several reasons. First, there were a limited number of replicate tests for most of the cable samples. Second, the electrical monitoring and thermal measurements were never made on the same cable, but rather on identical cables separated by a few centimeters in the tray or conduit. This was done to prevent interference between the thermocouple wire and the live cable conductors. Finally, as seen in Figure 4, the measured inner cable temperatures often increased dramatically just before the first electrical short because typically these types of cables ignite and fail electrically at about the same temperature (Nowlen and Wyant 2007b).

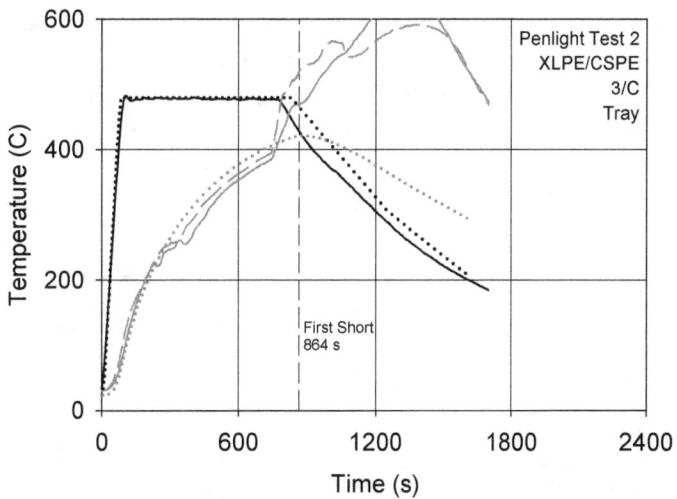

Figure 4. The results of Penlight Test 2 where it is observed that the inner cable temperatures (solid and dashed red lines) increase significantly about 1 min prior to the first recorded electrical short. The dotted lines are results of the THIEF model.

Table 2. Summary of the Sandia Penlight experiments that involved only single cables. Measurement uncertainties are reported in Nowlen and Wyant (2007b). Note that the thermoplastic cable (TEF/TEF) in Test 22 was tested like a thermoset.

Test	Cable No.	Cable Composition	Raceway	Exposing Temp. (°C)	Inner Cable Temperature (°C)		Time to First Short (s)
Thermosets							
PT-1	14	XLPE/CSPE	Tray	475	393	396	771
PT-2	14	XLPE/CSPE	Tray	470	393	410	864
PT-3	14	XLPE/CSPE	Tray	470	N/A	N/A	790
PT-7	14	XLPE/CSPE	Conduit	470	403	424	2334
PT-9	14	XLPE/CSPE	Air Drop	470	413	418	1531
PT-11	10	XLPE/CSPE	Tray	470	415	425	1225
PT-12	10	XLPE/CSPE	Tray	470	420	425	1273
PT-13	10	XLPE/CSPE	Tray	470	419	434	1198
PT-17	2	EPR/CPE	Tray	470	447	448	613
PT-18	9	SR/ Aramid Braid	Tray	700	N/A	N/A	DNF
PT-19	8	XLPO/XLPO	Tray	470	419	436	935
PT-20	3	XLPE/PVC	Tray	470	413	421	612
PT-22	12	TEF/TEF	Tray	470	382	384	445
PT-23	10	XLPE/CSPE	Conduit	470	425	429	1803
PT-24	10	XLPE/CSPE	Conduit	470	422	433	2006
PT-27	10	XLPE/CSPE	Air Drop	470	423	426	1356
PT-28	10	XLPE/CSPE	Air Drop	470	421	430	1314
PT-31	11	VITA-LINK	Tray	700	N/A	N/A	DNF
PT-62	13	XLPE/CSPE	Tray	470	N/A	N/A	502
PT-64	7	XLPE/CSPE	Tray	470	N/A	N/A	348
Thermoplastics							
PT-4	5	PVC/PVC	Tray	300	195	200	590
PT-5	5	PVC/PVC	Tray	300	211	213	766
PT-6	5	PVC/PVC	Tray	300	206	216	776
PT-8	5	PVC/PVC	Conduit	300	164	174	1245
PT-10	5	PVC/PVC	Air Drop	300	228	N/A	1173
PT-14	15	PE/PVC	Tray	300	237	238	1464
PT-15	15	PE/PVC	Tray	325	246	266	806
PT-16	15	PE/PVC	Tray	325	236	244	824
PT-21	1	PVC/PVC	Tray	300	196	229	560
PT-25	15	PE/PVC	Conduit	325	N/A	N/A	2924
PT-26	15	PE/PVC	Conduit	325	N/A	N/A	DNF
PT-29	15	PE/PVC	Air Drop	325	232	239	845
PT-30	15	PE/PVC	Air Drop	325	243	256	599
PT-63	6	PVC/PVC	Tray	325	205	208	333
PT-65	4	PVC/PVC	Tray	325	225	N/A	258

DNF Cable Did Not Fail electrically

3.3 Modeling Considerations

The thermally-induced electrical failure (THIEF) model has been incorporated into a computational fluid dynamics model called the Fire Dynamics Simulator or FDS (McGrattan *et al.* 2007). For the purpose of testing the THIEF model, the penlight apparatus was crudely modeled as a rectangular cavity, 0.6 m (2 ft) in length and 0.45 m (1.5 ft) in each transverse direction. The cable was modeled as a rectangular obstruction running the length of the cavity. Even though the obstruction is rectangular and much larger than the actual cable (to conform to the uniform, gas phase numerical grid), the one-dimensional heat conduction calculation within the "cable" is performed in cylindrical coordinates with appropriate dimensions. The rectangular obstruction is merely a convenient "target" for calculating the radiative and convective heat flux boundary condition for the solid phase heat conduction solver. The radiative heat flux is based on the solution of the radiative transport equation; the convective heat flux is based on empirical correlations. See McGrattan *et al.* (2007) for details.

It is certainly possible to model the penlight apparatus with greater spatial refinement, using FDS or another model of comparable capabilities. However, in practice the THIEF model is designed to work within a larger model that is to be used to assess the impact of a fire on an entire compartment. Resolution of such calculations is currently at best on the order of 5 cm (2 in) to 10 cm (4 in), making the cable essentially a "subgrid-scale" target. The ability of FDS to reproduce the thermal environment of the penlight apparatus is demonstrated in Figure 5. In Penlight Test 7 (PT-7), two cables were run through a conduit that was routed through the penlight. The conduit was modeled in FDS as a steel box with an assumed wall thickness of 4.9 mm, conductivity 45.8 W/m/K, density 7850 kg/m^3, specific heat 0.46 kJ/kg/K, and emissivity 0.85. The shroud temperature was reported by Nowlen and Wyant (2007b) as rising linearly to 478 °C. The prediction of the conduit temperature is within a few percentage points of the measurement, almost certainly within experimental uncertainty. However, the point of this exercise is not necessarily to validate the radiation and convection heat transfer algorithms within FDS, but rather to demonstrate that FDS can reproduce the thermal environment of the penlight apparatus. FDS is merely a convenient test bed for the cable failure algorithm.

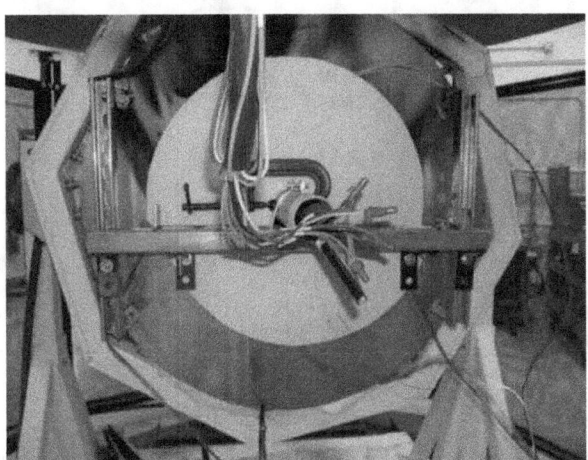
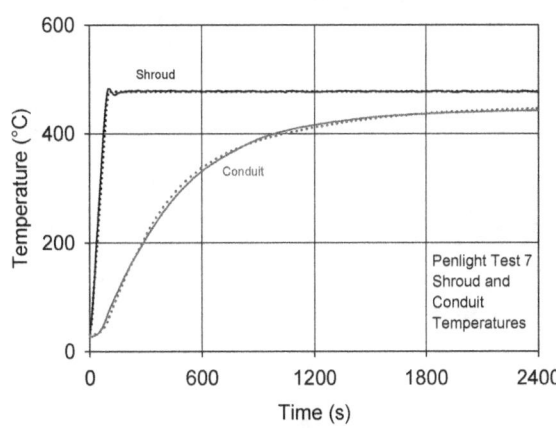

Figure 5. (Left) Photograph of a conduit within the penlight apparatus, with end caps installed, courtesy Frank Wyant, Sandia National Laboratories. (Right) Predicted (dotted line) and measured (solid line) temperatures of the conduit when exposed to the shroud temperature profile. This example is part of penlight experiment number 7.

3.4 Modeling Results

The THIEF model predictions of each of the 35 penlight experiments are presented in the following sections. Each experiment is summarized by a single graph, an example of which is shown in Figure 6.

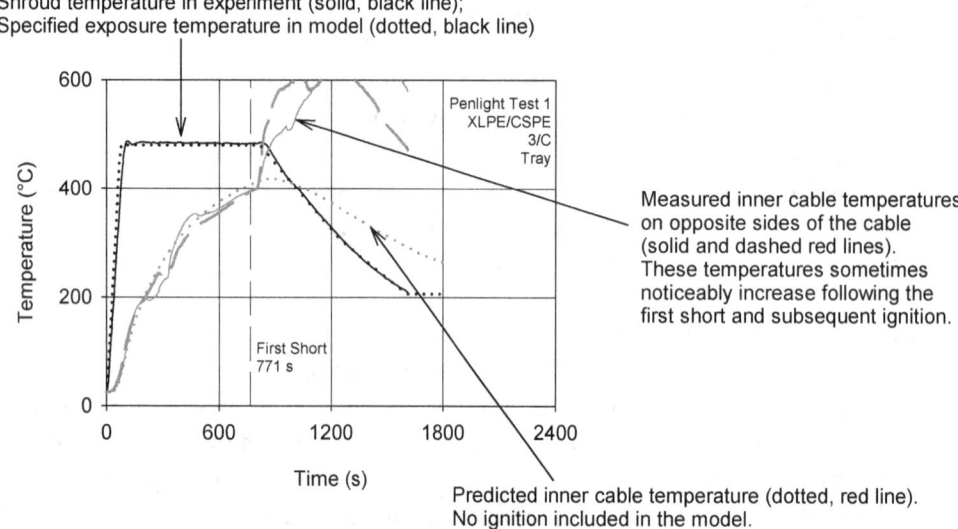

Figure 6. Key to graphs showing results of penlight tests. Note that the term "First Short" indicates the first observed electrical failure, regardless of the specific type. In the experiments, subsequent shorts were observed, but are not relevant here.

Thermoset Cables

Thermoset cables, in general, have been observed to short at higher temperatures than thermoplastics (Iqbal and Salley 2004). For this reason, the thermoset cables included in the CAROLFIRE program were exposed to higher temperatures in the Penlight Test Series. Even though the THIEF model does not distinguish between thermosets and thermoplastics, it is convenient to present the results according to this classification scheme.

Figure 7 through Figure 9 contain THIEF model predictions of the temperatures of various thermoset cables within the penlight apparatus.

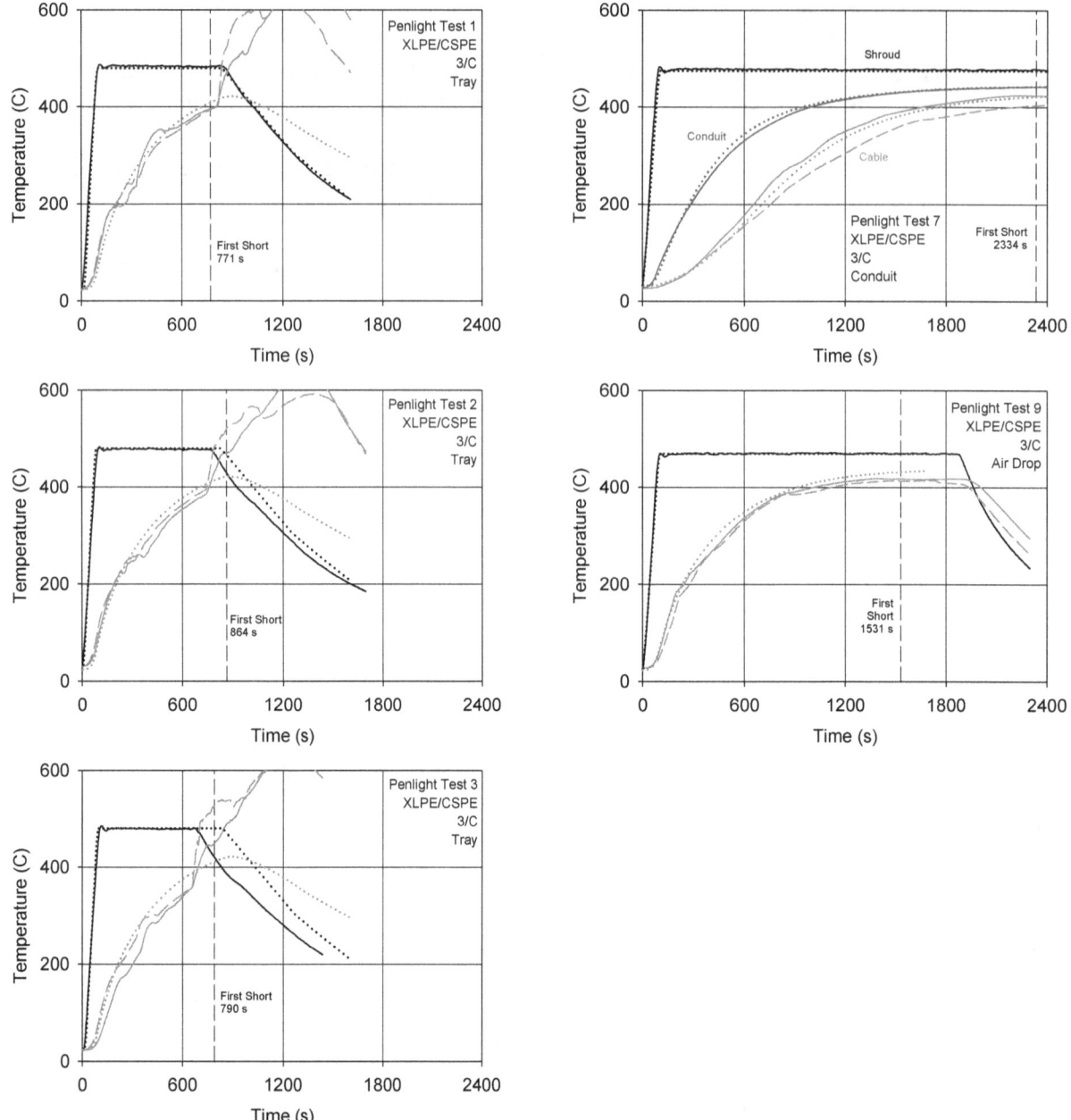

Figure 7. Results for Cable #14 (XLPE/CSPE, 3 conductor). The upper-most black curves depict the measured (solid) and specified (dotted) shroud temperature. The solid and dashed red lines are the measured temperatures inside the cable jacket, at opposite sides. The dotted red curve is the predicted inner cable temperature. The vertical dashed line indicates the first observed electrical short in the experiment. For Test 7, the solid blue and dotted blue curves represent the measured and predicted conduit temperatures, respectively.

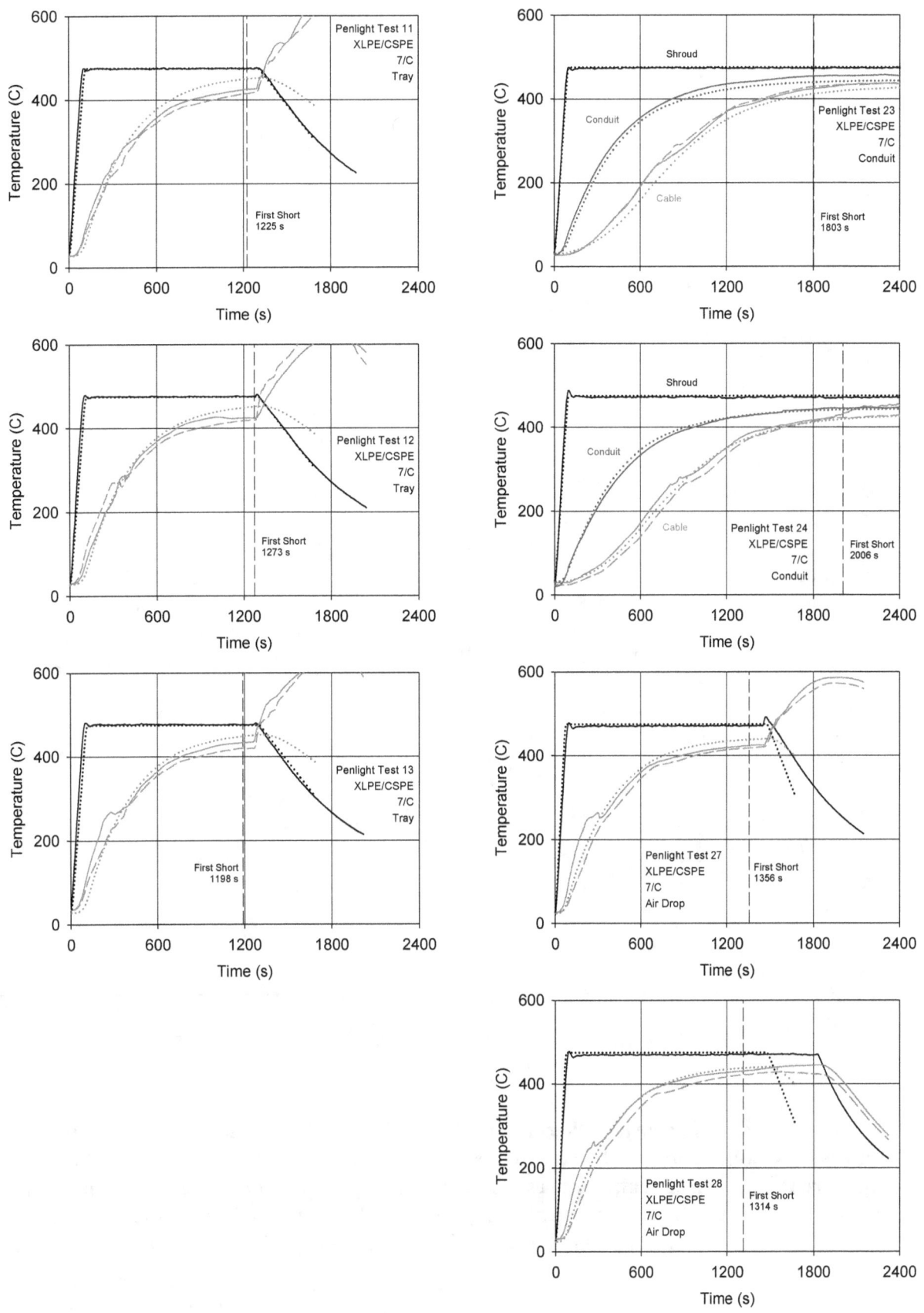

Figure 8. Results for Cable #10 (XLPE/CSPE, 7 conductor). The color scheme is the same as in Figure 7.

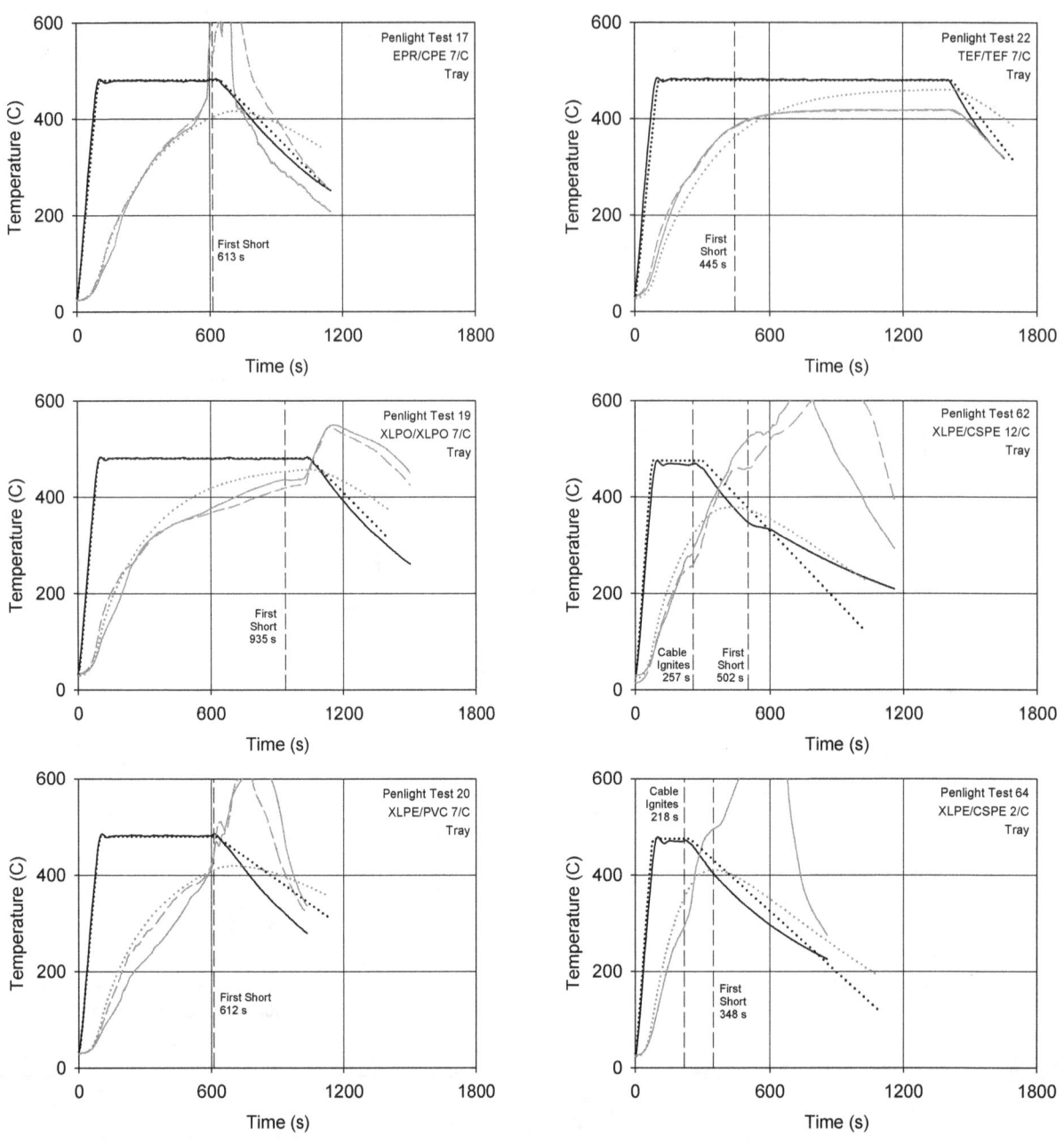

Figure 9. Results for various thermoset cables. The upper-most black curves depict the measured (solid) and specified (dotted) shroud temperature. The solid and dashed red lines are the measured temperatures inside the cable jacket, at opposite sides. The dotted red curve is the predicted inner cable temperature. The vertical dashed line indicates the first observed electrical short in the experiment.

Thermoplastic Cables

The shroud temperature for the single cable thermoplastic penlight tests was set to just over 300 °C (572 °F). For the THIEF model predictions, the same thermal conductivity and specific heats that are used for the thermosets are also used for the thermoplastics. Thus, no distinction is made in the THIEF model between thermosets and thermoplastics.

Figure 10 through Figure 12 contain THIEF model predictions of the temperatures of various thermoplastic cables within the penlight apparatus.

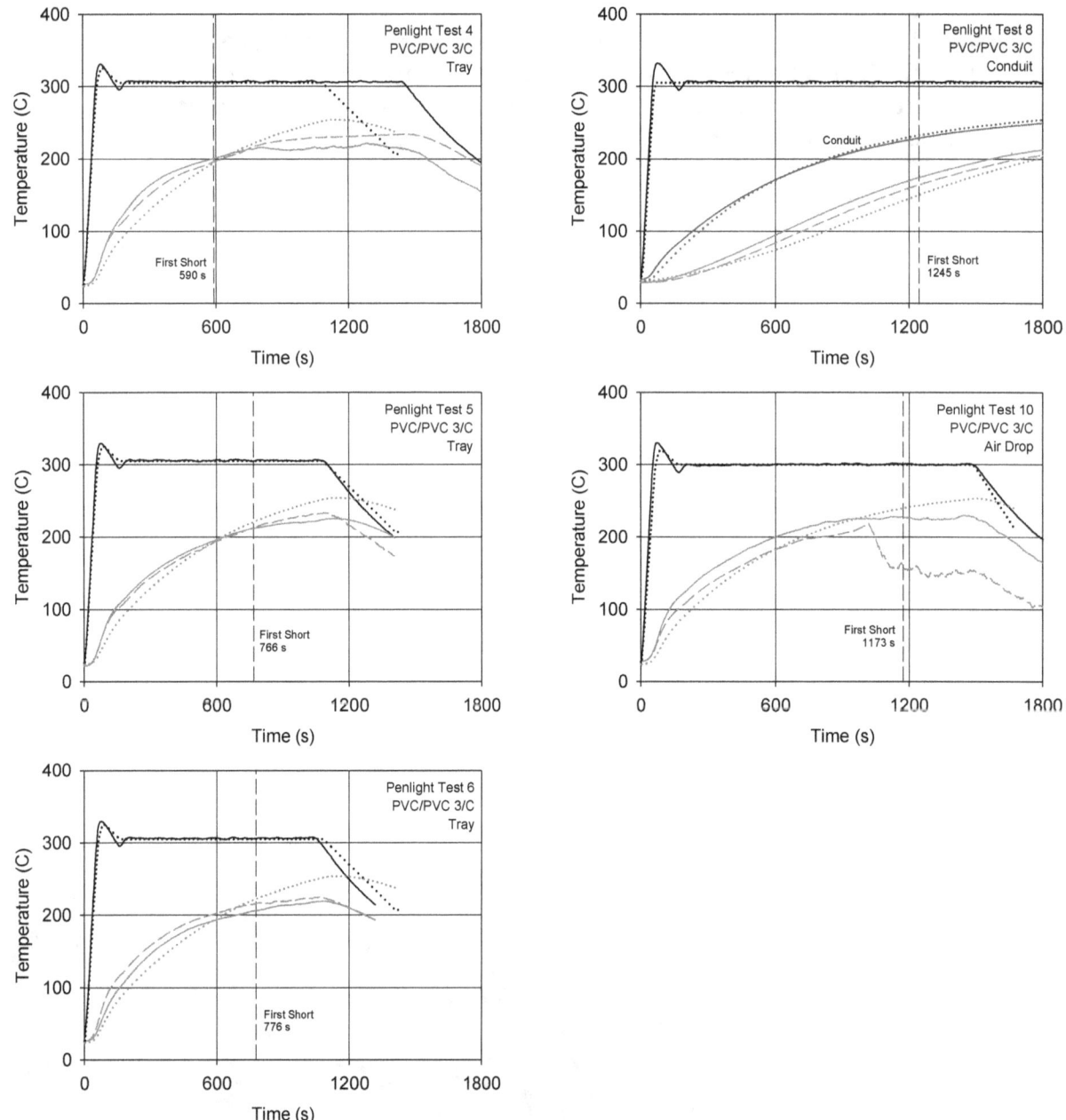

Figure 10. Results for Cable #5 (PVC/PVC, 3 conductor). The upper-most black curves depict the measured (solid) and specified (dotted) shroud temperature. The solid and dashed red lines are the measured temperatures inside the cable jacket, at opposite sides. The dotted red curve is the predicted inner cable temperature. The vertical dashed line indicates the first observed electrical short in the experiment. For Test 8, the solid blue and dotted blue curves represent the measured and predicted conduit temperatures, respectively.

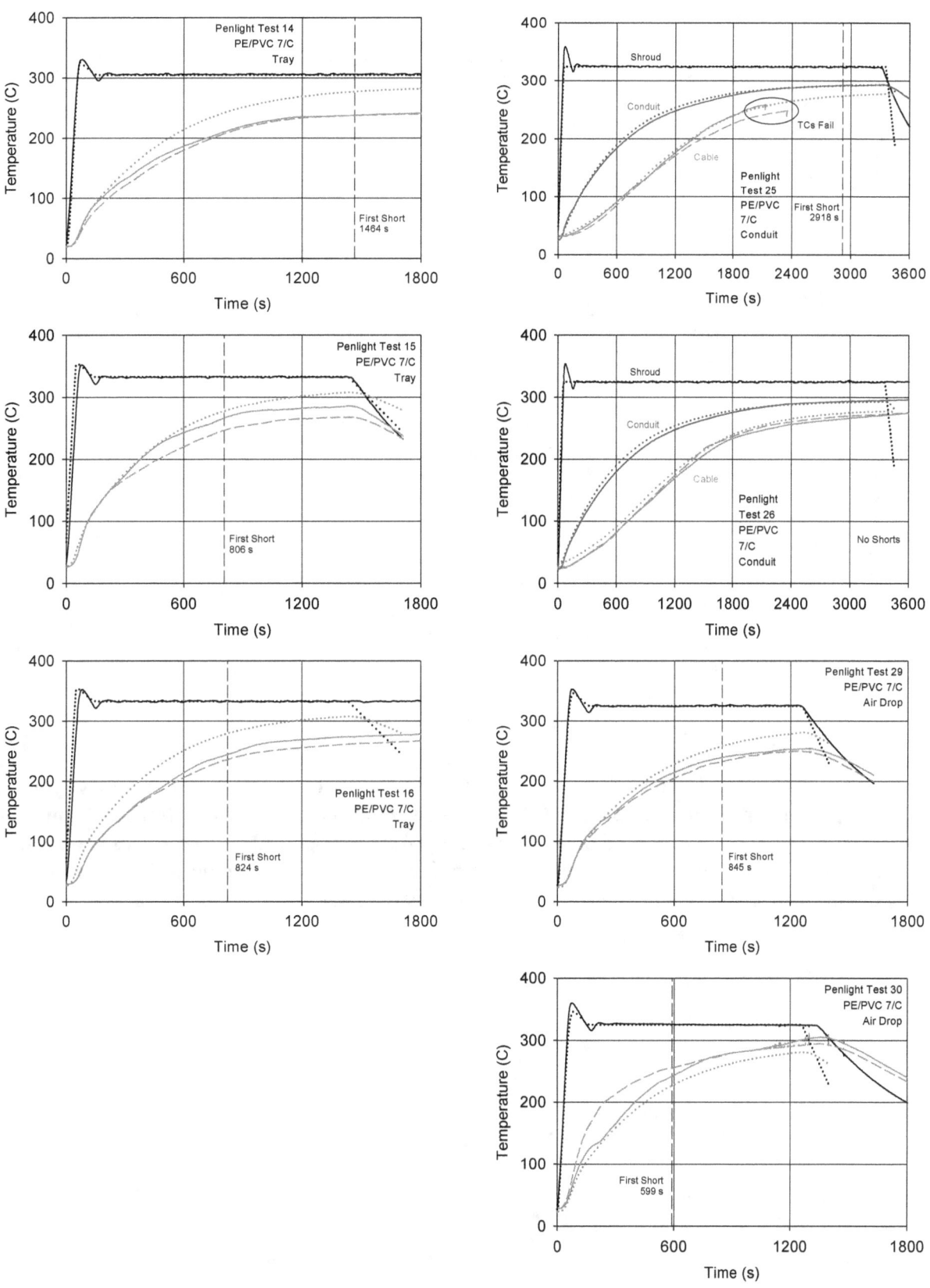

Figure 11. Results for Cable #15 (PE/PVC, 7 conductor). The color scheme is the same as in Figure 10.

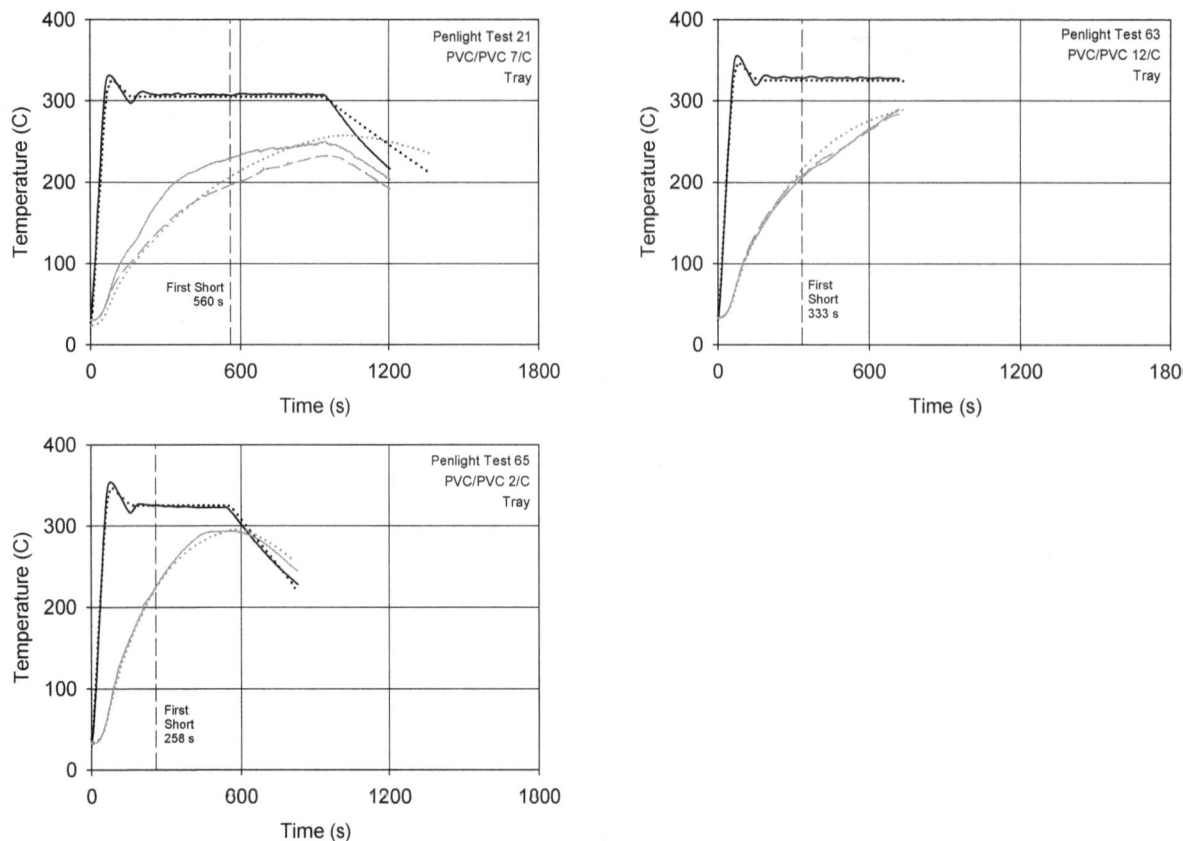

Figure 12. Results for 2, 7, and 12 conductor PVC/PVC (thermoplastic) cables. The upper-most black curves depict the measured (solid) and specified (dotted) shroud temperature. The solid and dashed red lines are the measured temperatures inside the cable jacket, at opposite sides. The dotted red curve is the predicted inner cable temperature. The vertical dashed line indicates the first observed electrical short in the experiment.

Special Cables

Two of the fifteen cable samples were observed during preliminary experiments to withstand significantly higher temperatures than the others without any observed electrical shorting. Thus, these two cables were tested at higher temperatures during the Penlight test series. Figure 13 presents the results of the experiments and THIEF model predictions. Note that the cables were not observed to fail electrically during the experiments, even though they did ignite and burn. This same behavior was observed during the Intermediate Scale experiments, but the cables did fail electrically when a water spray was applied. See Nowlen and Wyant (2007b) for details.

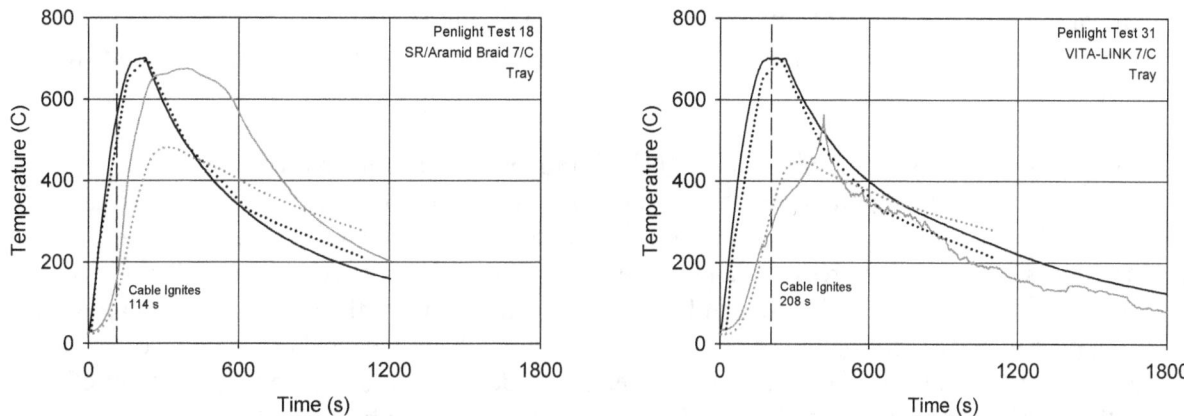

Figure 13. Results for Cable #9 (Silicone Rubber/Aramid Braid, 7 conductor) and Cable #11 (VITA-LINK, 7 conductor). The upper-most black curves depict the measured (solid) and model-specified (dotted) shroud temperature. The solid red line is the measured temperature inside the cable jacket. The dotted red line is the predicted inner cable temperature. Ignition and burning are not included in the THIEF model, and electrical shorting did not occur in these two experiments. Note that neither cable failed electrically until the introduction of a water spray. See Nowlen and Wyant (2007b) for details.

3.5 Summary of Penlight Analysis

For the purpose of evaluating the THIEF model, it is sufficient to simply choose a "threshold temperature" for each class of cable to serve as a surrogate for a true failure temperature that would have to be determined from a more extensive set of measurements. For this exercise, 400 °C (752 °F) was chosen for the thermosets; 200 °C (392 °F) for the thermoplastics. Figure 14 compares the predicted time to the threshold temperature versus the measured times for the 35 single cable tests chosen from the Penlight series. The data for this graph are included in Table 3. There are two inner-cable measurements considered, made on opposite sides of the cable, just beneath the jacket. In all, 66 THIEF model predictions of time to "threshold" temperature were compared to the measured counterparts (in 4 tests, the Point B measurement was not made). The THIEF model under-predicted the times by 3 %, on average, and the standard deviation was 20 %.

Overall, there is only a slight bias in the THIEF model towards under-predicting the time to reach the threshold temperature in the ideal environment of the Penlight apparatus. To test whether the model has a particular bias related to the various cable properties, the relative differences[6] between the predicted and measured "failure" times were plotted as functions of the various cable properties and shown in Figure 15. From the plots, the accuracy of the model does not appear to be related to the various bulk properties of the cable, as there is no discernable pattern in the various graphs. If, for example, the model were to over-predict the "failure" time for thin cables, one would expect to see that reflected in the graph.

---

[6] The relative difference was calculated as the difference between the predicted and the measured time divided by the measured time. A positive value of the relative difference means that the model over-predicted that particular threshold time.

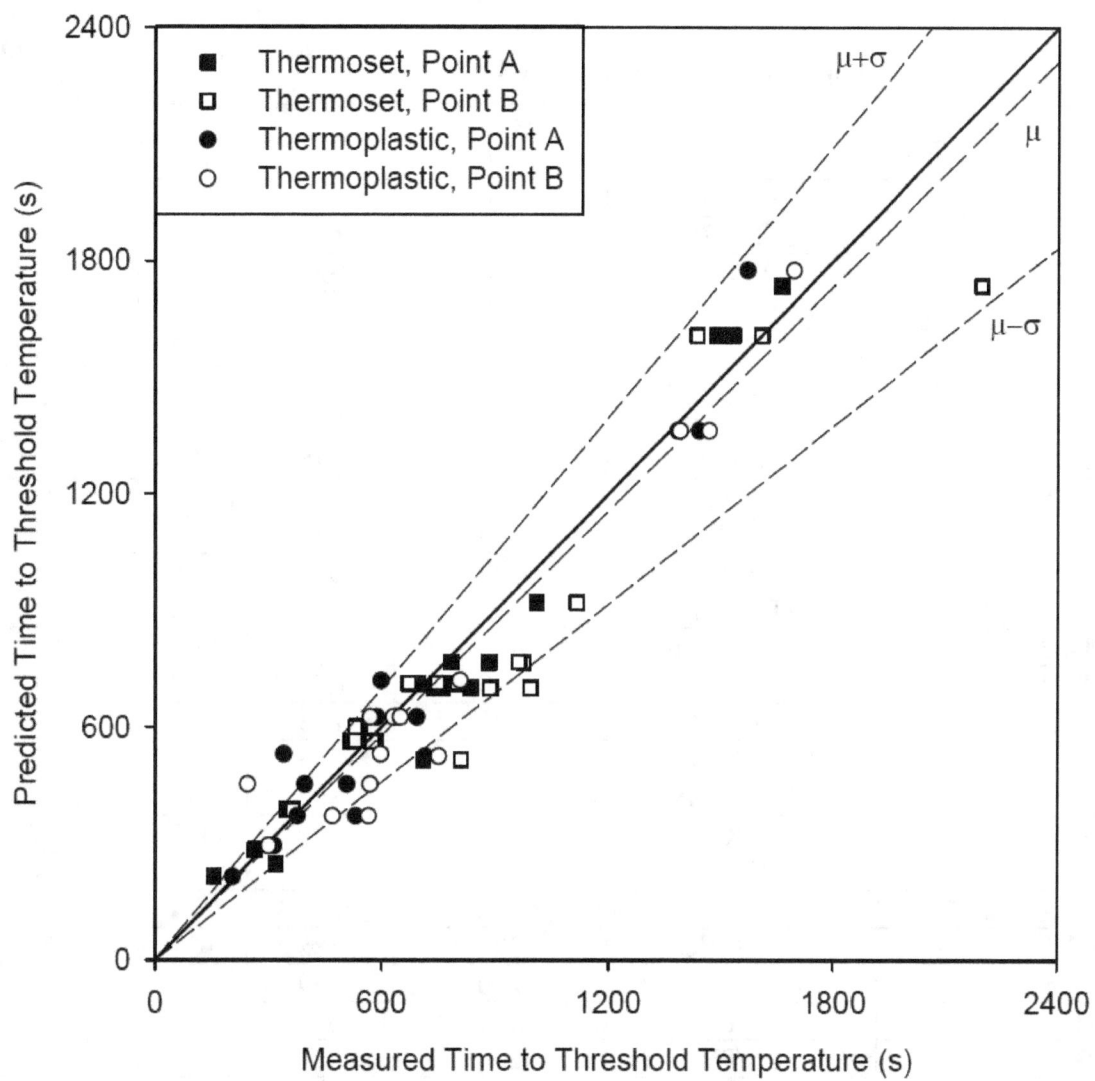

Figure 14. Comparison of THIEF model and experiment for the Penlight series. Point A and B refer to the thermocouple measurements made on opposite sides of the cable, just under the jacket. The "Threshold Temperature" is 400 °C (752 °F) for thermoset cables and 200 °C (392 °F) for thermoplastics. The dashed lines represent the average (-3 %) and standard deviation (20 %) of the data.

Table 3. Summary of the THIEF Model Predictions of the Penlight Experiments.

Test	Cable No.	Cable Composition (Insulation/Jacket)	Threshold Temperature (°C)	Measured Time to Threshold (s)		Predicted Time to Threshold (s)
Thermosets						
PT-1	14	XLPE/CSPE	400	800	800	712
PT-2	14	XLPE/CSPE	400	761	747	712
PT-3	14	XLPE/CSPE	400	694	671	712
PT-7	14	XLPE/CSPE	400	1660	2196	1735
PT-9	14	XLPE/CSPE	400	1009	1115	920
PT-11	10	XLPE/CSPE	400	835	993	700
PT-12	10	XLPE/CSPE	400	740	887	700
PT-13	10	XLPE/CSPE	400	761	885	700
PT-17	2	EPR/CPE	400	545	533	600
PT-18	9	SR/ Aramid Braid	400	157	--	216
PT-19	8	XLPO/XLPO	400	710	810	516
PT-20	3	XLPE/PVC	400	585	575	562
PT-22	12	TEF/TEF	400	518	530	564
PT-23	10	XLPE/CSPE	400	1488	1434	1608
PT-24	10	XLPE/CSPE	400	1532	1608	1608
PT-27	10	XLPE/CSPE	400	883	975	766
PT-28	10	XLPE/CSPE	400	785	961	768
PT-31	11	VITA-LINK	200	322	--	247
PT-62	13	XLPE/CSPE	400	349	365	387
PT-64	7	XLPE/CSPE	400	265	--	284
Thermoplastics						
PT-4	5	PVC/PVC	200	588	631	625
PT-5	5	PVC/PVC	200	639	649	625
PT-6	5	PVC/PVC	200	693	571	625
PT-8	5	PVC/PVC	200	1570	1692	1775
PT-10	5	PVC/PVC	200	599	807	720
PT-14	15	PE/PVC	200	715	750	525
PT-15	15	PE/PVC	200	378	471	371
PT-16	15	PE/PVC	200	532	566	371
PT-21	1	PVC/PVC	200	342	598	530
PT-25	15	PE/PVC	200	1382	1467	1362
PT-26	15	PE/PVC	200	1439	1388	1362
PT-29	15	PE/PVC	200	509	570	453
PT-30	15	PE/PVC	200	397	246	453
PT-63	6	PVC/PVC	200	316	302	294
PT-65	4	PVC/PVC	400	206	--	215

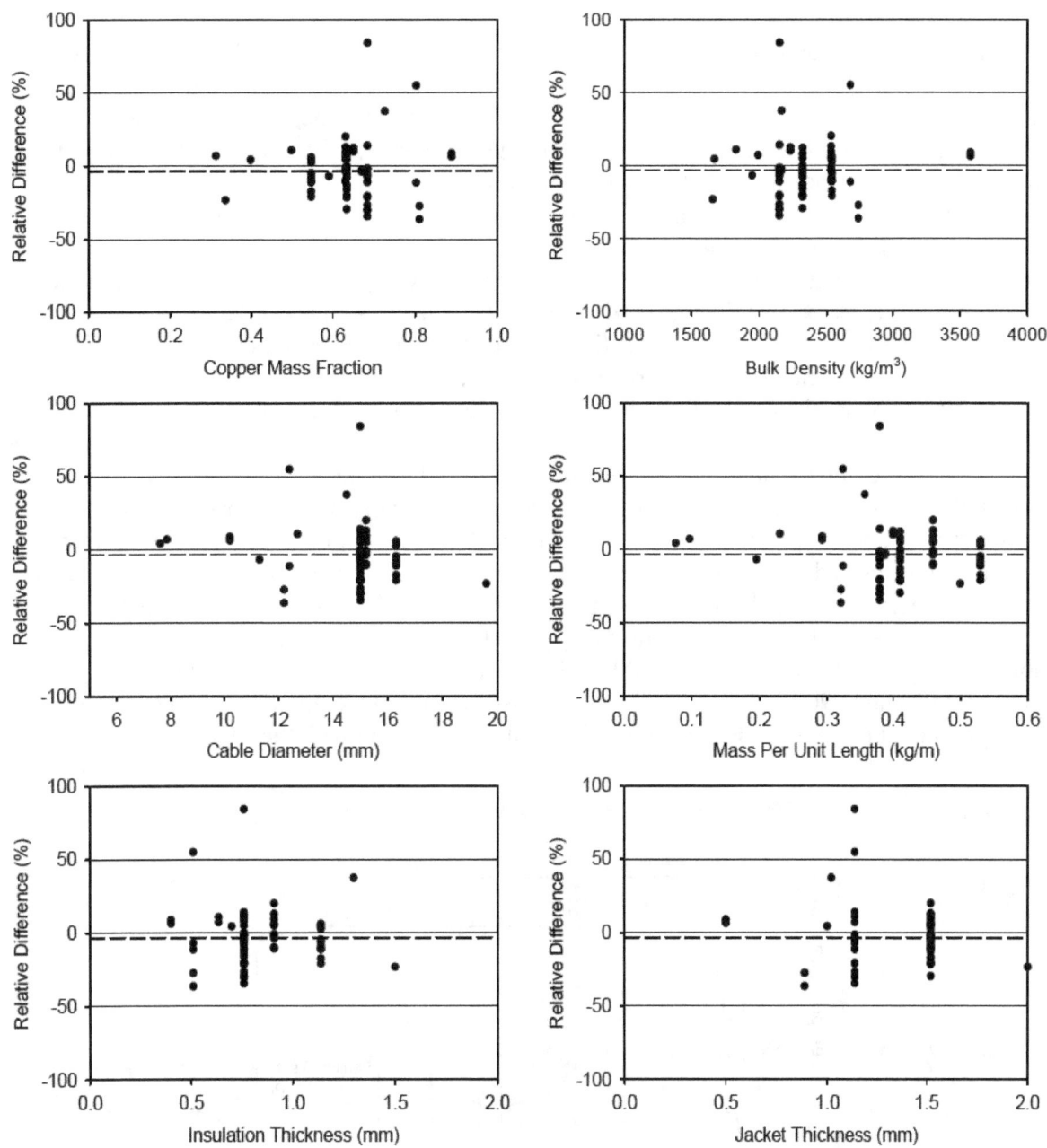

Figure 15. Sensitivity of the THIEF model predictions to various cable properties. The Relative Difference refers to the predicted vs measured times to reach the threshold temperature. The horizontal dashed line is the average Relative Difference for all the Penlight comparisons.

3.6 Why Does the THIEF Model Work?

Given the complexity and variety of composition among the 15 different cable types that were tested in the CAROLFIRE Penlight series, it is reasonable to ask why the simplest of mathematical models could have predicted the results of the experiments so well. The answer lies in the fact that simple "lumped parameter" models often take advantage of off-setting errors. That is, the errors associated with the various modeling assumptions often act to partially cancel each other out. The most important assumption in the THIEF model is that the cable is a homogenous cylinder with temperature-independent thermal properties. The thermal conductivity and specific heat are fixed for all cable types, with values typical of both thermoset and thermoplastic materials (Hamins et al. 2006). The density is an effective value, based on the overall mass per unit length and cross sectional area of the cable. Even though the mass fraction of copper in the cables tested is in the range of 0.3 to 0.9, its contribution is only through the bulk density. Neither the conductivity nor specific heat of copper plays a role in the model.

Suppose, however, that a slightly more detailed model of the cable is formulated. Assume that instead of a homogenous cylinder, the cable is idealized as a polymeric/copper mixture, surrounded by a purely polymeric jacket. The thermal properties of the mixture are calculated based on the measured mass and volume fractions of the respective materials within a particular cable. The thermal conductivity, specific heat and density of the polymer are assumed to be 0.2 W/m/K, 1.5 kJ/kg/K, and 1380 kg/m^3, respectively, all based on measurements of several cables and reported in Hamins et al. (2006). The properties of copper are 400 W/m/K, 0.385 kJ/kg/K, and 8960 kg/m^3, respectively (Weast 1982). The temperature prediction of this two-layer model is shown in Figure 16 for Penlight Test 1, along with the THIEF prediction.

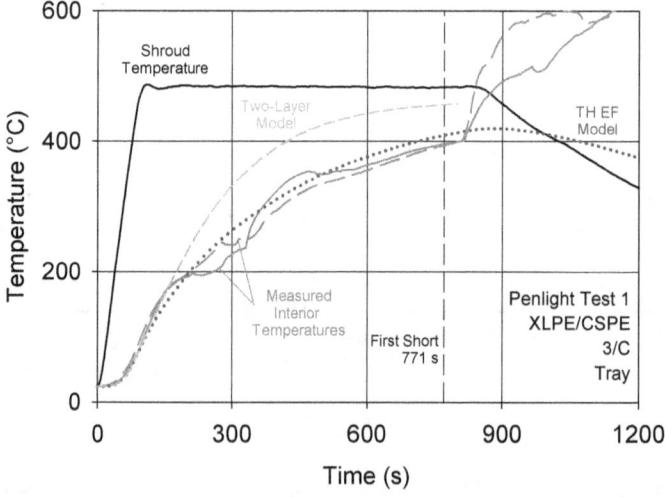

Figure 16. Results of an alternative thermal model (short dashes) and the THIEF model (dotted line), compared to measurements (solid and long dashed lines).

The prediction of the more detailed model is, at least in this case, less accurate than the THIEF model. Why? Consider the noticeable shift in the two measured interior temperatures at about 200 s and 200 °C. Neither the THIEF nor the two-layer model captures this abrupt shift in temperature because neither model accounts for decomposition reactions or evaporation which inevitably occurs as the polymers heat up and produce this characteristic pattern in the temperature plot. The two-layer model is a better description of the cable structure and composition, as is apparent from its closer match to the measurement in the first 200 s. However, the rise of the measured temperatures inside the cable noticeably slows at this point, most likely caused by some endothermic process that draws energy from the system and slows the temperature rise. The THIEF model uses the specific heat of the polymer (1.5 kJ/kg/K) throughout the cable cross section, whereas the two-layer model uses this value only for the jacket layer, and 0.75 kJ/kg/K for the interior mixture. Also, the two-layer model uses the density of the polymer (1380 kg/m^3) for the jacket layer, whereas the THIEF model uses the bulk density of the entire cable (2538 kg/m^3). In short, the THIEF model uses an upper-bound estimate for the thermal inertia (density times specific heat) of the cable, but at the same time neglects the endothermic polymeric decomposition reactions. The two-layer model better estimates the thermal inertia and overall cable construction, but it too neglects the reactions. THIEF works because of the off-setting errors – its upper-bound estimate of the thermal inertia counteracts, to some degree, its neglect of the endothermic reactions. The two-layer model has no mechanism to counteract its neglect of the reactions.

A natural question to ask is why not include chemical reactions in the model? Indeed, as part of the CAROLFIRE program, a detailed model of a cable -- including multi-component kinetics, temperature-dependent thermal properties, and multi-dimensional heat conduction -- was developed at the University of Maryland (Chourio 2007). However, the detailed model could not be used to predict the measured temperatures because of a lack of property data:

> "Given the scarcity of data and the lack of consistency in the characterization of the kinetic parameters, it is not possible to assess the endurance limits for this model at this time."

As a general rule, a more detailed model requires more detailed inputs. If the thermal and chemical properties of the polymers are not available, simplifications have to be made. Furthermore, even if the property data were provided or measured, the more detailed model would not provide more accurate results, at least for single cables, because of the natural variability of the measured failure times. Consider that the replicate Penlight Tests 1, 2 and 3 had measured failure times of 771 s, 864 s and 790 s, respectively; and that replicate Penlight Tests 4, 5 and 6 had measured failure times of 590 s, 766 s and 776 s, respectively.

4 INTERMEDIATE SCALE EXPERIMENTS

4.1 Experimental Description

Following the Penlight Test Series, cables in various configurations were exposed to a realistic fire environment in what are referred to as the "Intermediate Scale Tests." In these experiments, an ethylene gas burner was centered underneath a 3.6 m (12 ft) by 2.4 m (8 ft) by 1.3 m (4 ft) deep enclosure constructed of gypsum board that was suspended about 2 m (6 ft) above the floor (see Figure 17). Cable trays, conduits, and "air drop" cables were exposed to fires ranging from 250 kW to 350 kW. Some cables were monitored for electrical response; some for thermal response.

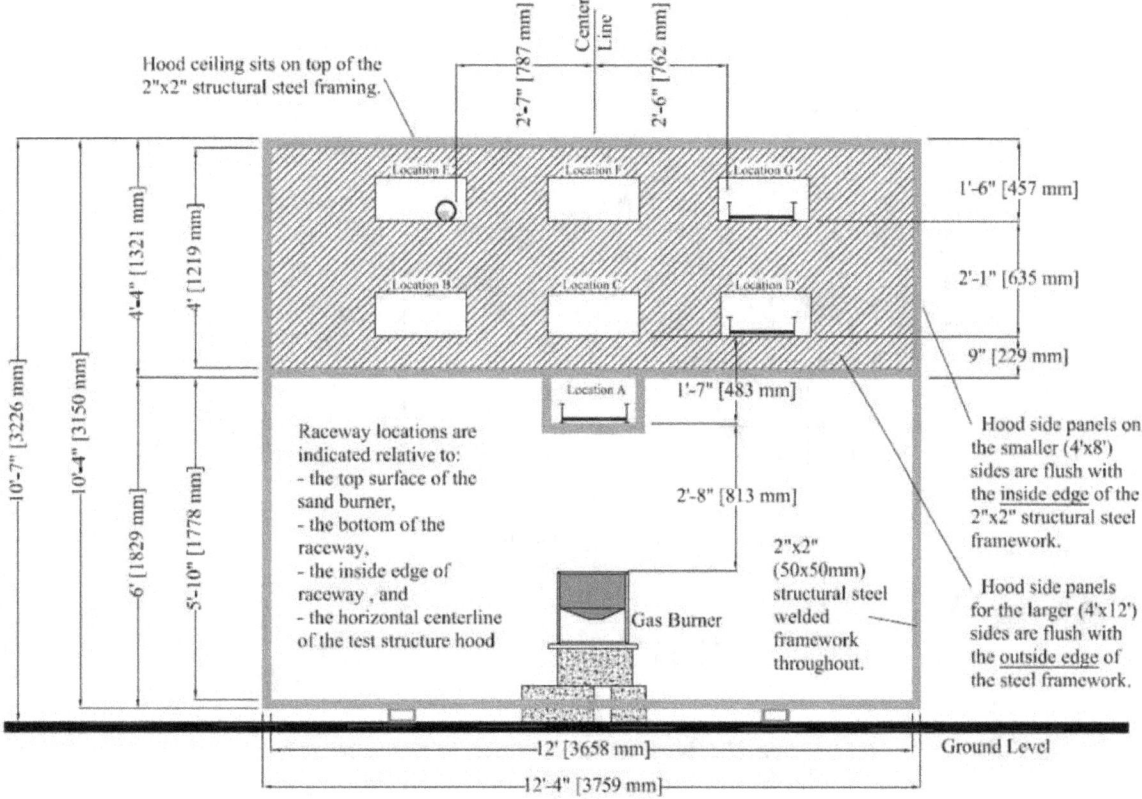

Figure 17. Side view of the Intermediate Scale Test rig. Courtesy Sandia National Laboratories.

4.2 Modeling Considerations

Initially, these experiments were simulated using the Fire Dynamics Simulator, and predictions were made for the cable failure times based on the THIEF model that was embedded within FDS. However, because the heat release rate of the burning cables was not measured, it was not possible to predict accurately the gas temperatures in the vicinity of the various cable trays. Without sound predictions of the enclosure gas temperatures, it was not possible to test the

THIEF model within a larger simulation of the entire compartment.

Fortunately, the measured gas temperature from the experiments served as appropriate "exposing" temperatures for testing the cable failure algorithm. In other words, the measured gas temperatures were used much like the "shroud" temperatures in the Penlight test series. Thus, the THIEF model was used to predict the inner temperature of every cable with an embedded thermocouple and a gas temperature measurement in its vicinity (usually above or below the tray). In the case of conduits, the measured conduit temperature was used as the "exposing" temperature. Figure 18 displays the result of a typical comparison of model and measurement.

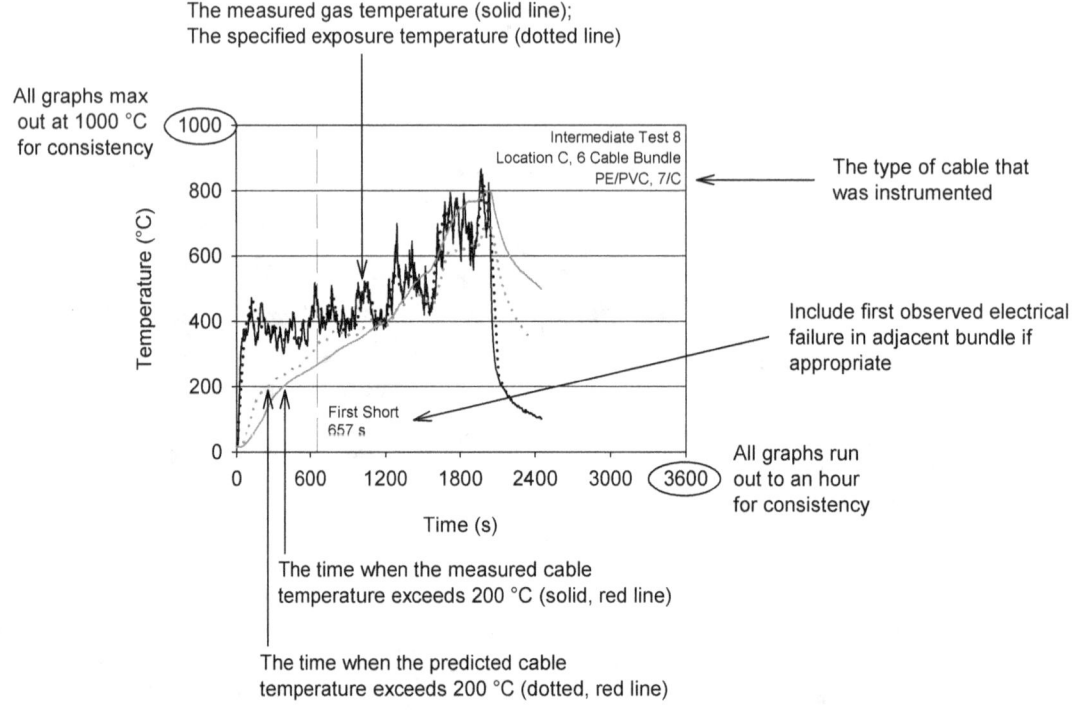

Figure 18. Key to Intermediate Scale graphs.

Note that the graph in Figure 18 includes a vertical dashed line indicating when the first electrical short was observed in the experiment. This is only included for situations when comparable cables were instrumented for both thermal and electrical response. This was not always the case, and it is noted on the plots when there was no electrical monitoring present.

Even in cases where two equivalent cables were monitored for thermal and electrical response, it is not always appropriate to link the measured temperature with electrical failure. For example, Figure 19 shows the measured and predicted inner temperature (TC-1) of a single cable within a 6-cable bundle. Given the configuration, it is likely that the first short occurs at the bottom of Cable E, not the top, where the temperature is measured. The THIEF model cannot distinguish the top and bottom of the cable because it only accounts for heat transfer in the radial direction.

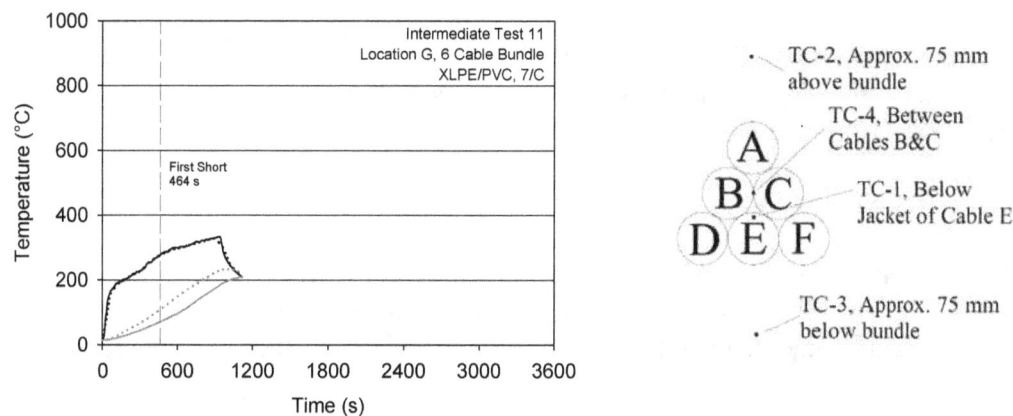

Figure 19. Example of an experiment where the temperature measurement (TC-1) may not coincide with the location of the first electrical short.

4.3 Model Results

The following sections describe the results of the THIEF predictions for the various cable configurations, including single cables within trays, cables within conduit, random fill trays, six cable bundles, 12 cable bundles and "air drops."

Single Cables within Trays

In several Intermediate Scale Tests, single cables were laid out in trays in pairs – one instrumented for electrical response, the other for thermal. The plots on the following page show comparisons of the cable inner temperature measurements and the corresponding predictions by the THIEF model.

Note that for the single cable cases, *and only the single cable cases*, the measured gas temperature was modified to account for the influence of the heat flux from the fire directly on the cables. Because the direct impingement of thermal radiation from the fire provided a significant fraction of the total heat flux early in the test, the measured gas temperature, T_{gas}, was modified to account for the additional heat flux from the fire, $\dot{q}''$

$$\sigma\left(T_{eff}^4 - T_\infty^4\right) = \sigma\left(T_{gas}^4 - T_\infty^4\right) + \dot{q}'' \qquad (4.1)$$

The heat flux from the fire was estimated using a point source approximation, with the origin centered above the burner a distance of half the flame height as calculated using Heskestad's correlation (Iqbal and Salley 2004). The "effective" gas temperature, T_{eff}, was used as input in the THIEF model as the exposing temperature, rather than the actual measured gas temperature. In all other cases, the instrumented cables were shielded from the fire's direct thermal radiation by either a conduit or surrounding cables within a bundle, and no correction to the measured gas temperature was needed.

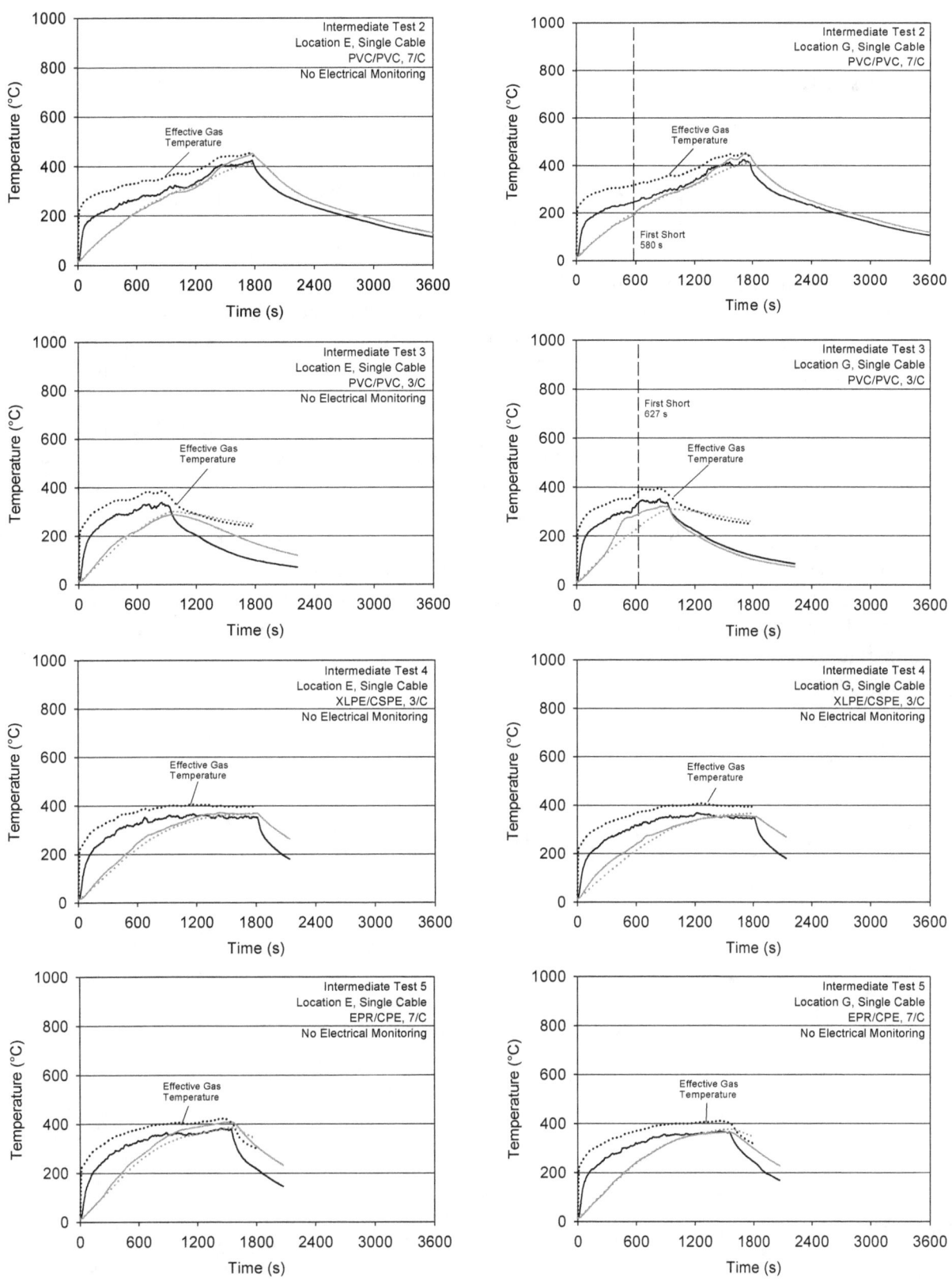

Figure 20. Summary of results for single, isolated cables in the Intermediate scale tests. Black indicates the exposing temperature; red the cable. Solid lines for experiment, dotted for model.

Cables in Conduits

In 10 instances during the Intermediate Scale Test Series, cables were routed through a standard heavy wall, galvanized steel conduit in bundles similar to that shown below. The cable labeled with the number 1 was instrumented with a thermocouple inside its jacket, while the lettered cables were instrumented for electrical response. For the purpose of testing the THIEF model, the measured conduit temperature itself was used as the exposing temperature, and the measured temperature of Cable 1 was compared to that predicted by the model.

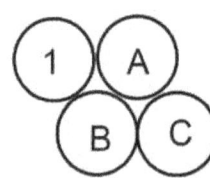

The results for the conduit at Location E (see Figure 17 for locations) are shown in Figure 21; the results for Location D and G in Figure 22. These results are most similar to the Penlights because the conduit is very much like the "shroud" of the penlight apparatus in that it distributes the radiative flux uniformly about the instrumented cable. Thus, the one-dimensional heat conduction assumption holds. Note that in the Intermediate Scale Tests, no gas phase measurements were taken outside the conduit, and no attempt was made to predict the conduit temperature, as was the case in the Penlight Tests. Rather, the conduit temperature was taken as the *specified* exposing temperature. From the results shown on the following pages, it appears that the conduit provided a fairly uniform thermal exposure for the cables inside, much like the cylindrical shroud in the penlight apparatus.

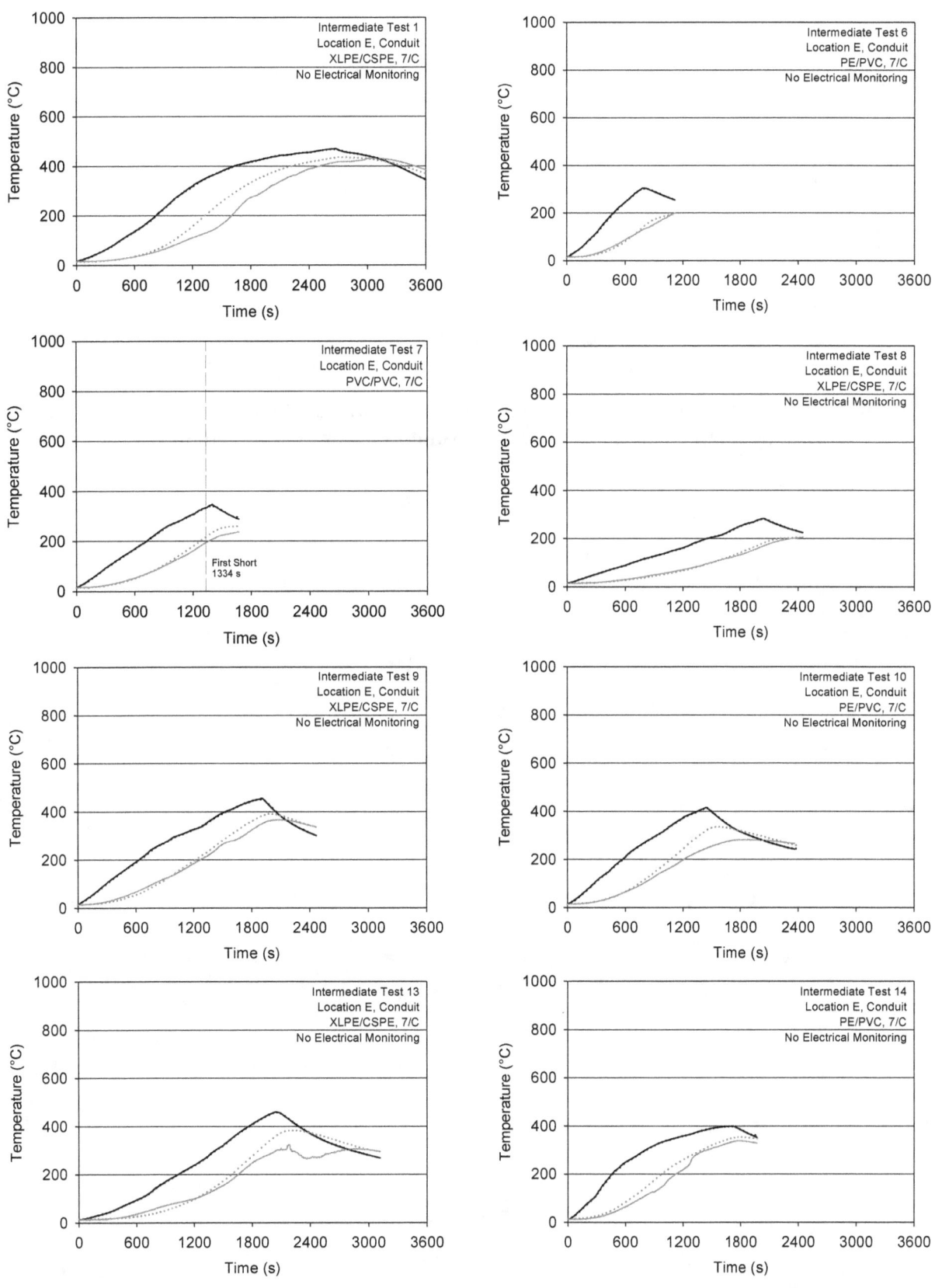

Figure 21. Summary of results for the cables in the conduit at Location E. Black indicates the exposing temperature; red the cable. Solid lines for experiment, dotted for model.

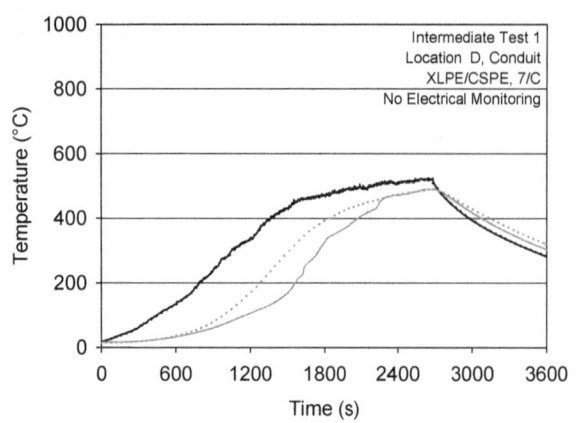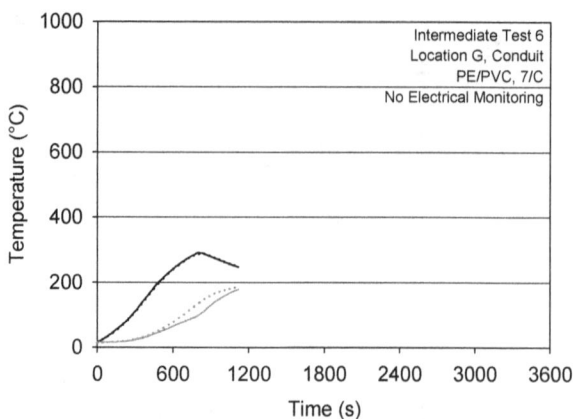

Figure 22. Summary of results for cables in conduits at Location D and G. Black indicates the exposing temperature; red the cable. Solid lines for experiment, dotted for model.

Random Fill Cable Trays

In a few of the Intermediate Scale Tests, Trays A and C contained a random mixture of cables, a few of which were instrumented with a thermocouple beneath the cable jacket. Figures 22 and 23 below show the two different configurations. The first is the composition of Tray A during Test 1, the second is that of Trays A and C during Tests 13 and 14.

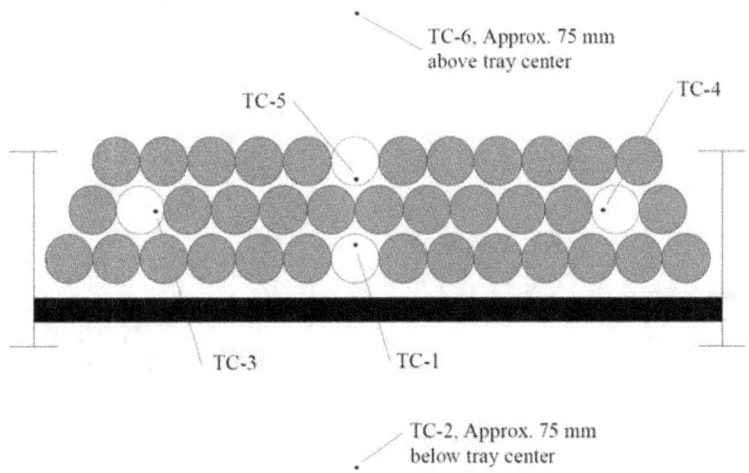

Figure 23. Cable tray configuration for Intermediate Test 1.

For Test 1, the exposing temperature corresponding to the instrumented cable in the bottom row (TC-1) was measured by TC-2. The exposing temperature for the top row cable (TC-5) was measured by TC-6. TC-3 and TC-4 were not used.

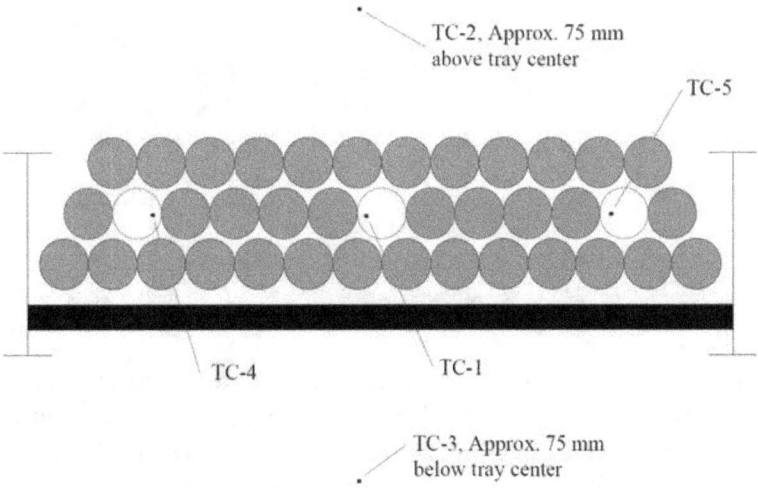

Figure 24. Cable tray configuration for Intermediate Tests 13 and 14.

In Tests 13 and 14, both Tray A and C were filled with random cables. The instrumented cable in the middle of the center row (TC-1) corresponded to an exposing temperature measured below the bundle (TC-3). TC-4 and TC-5 were not used.

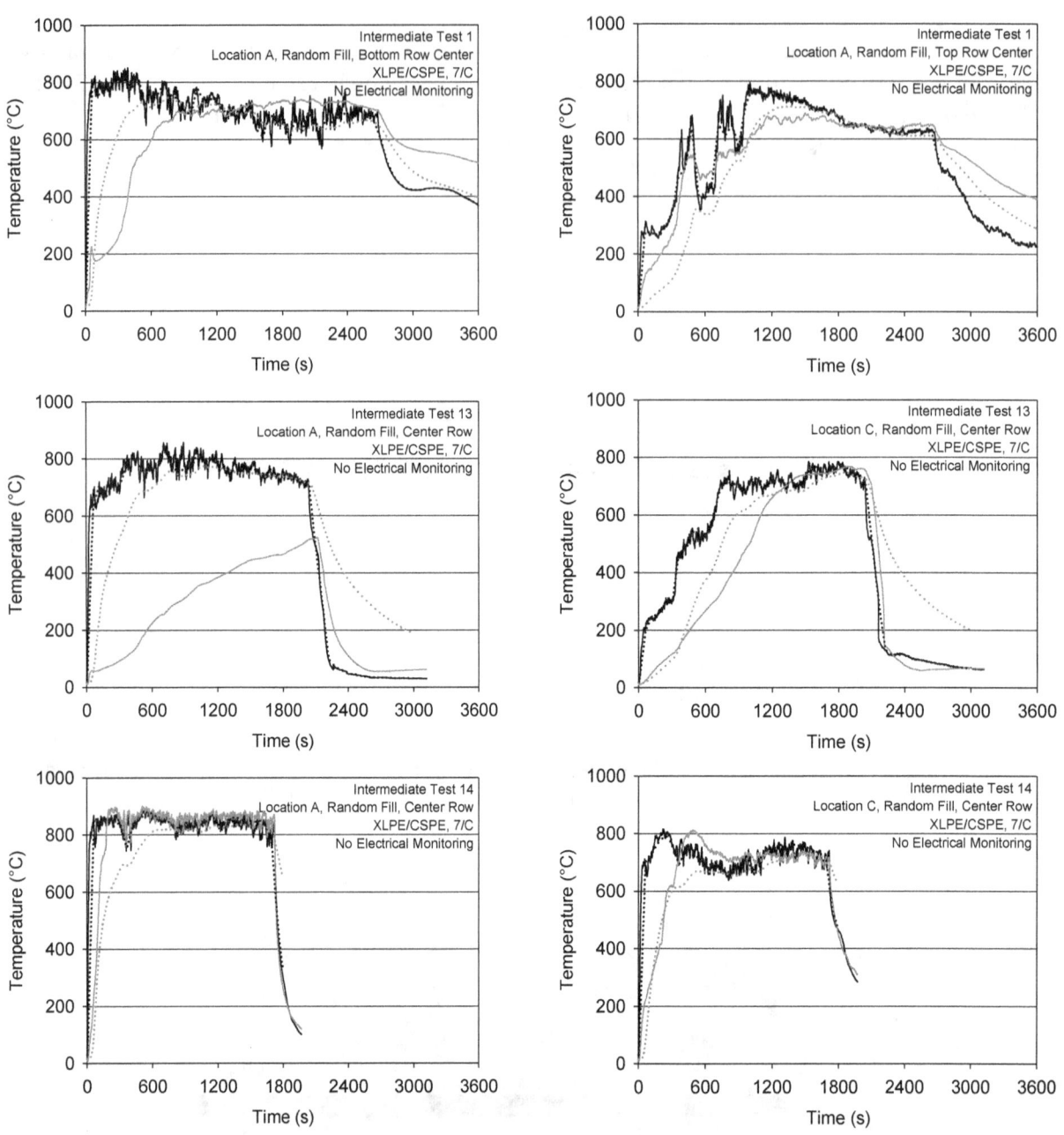

Figure 25. Summary of results for trays filled with random cable. Black indicates the exposing temperature; red the cable. Solid lines for experiment, dotted for model. The measured temperature at Location A in Test 13 (middle left graph) is inconsistent with that at Location C (middle right graph).

Air Drops

There were 5 instances during the Intermediate Scale Tests when a single or small bundle of cables were configured as an "air drop;" that is, the cables were not supported by a tray or conduit. The results are shown in Figure 26. In three cases, an instrumented bundle of cables was "dropped" from Tray C down to Tray A. Both trays were directly in the fire plume. In the other two cases, the cables were suspended wall to wall at Location E.

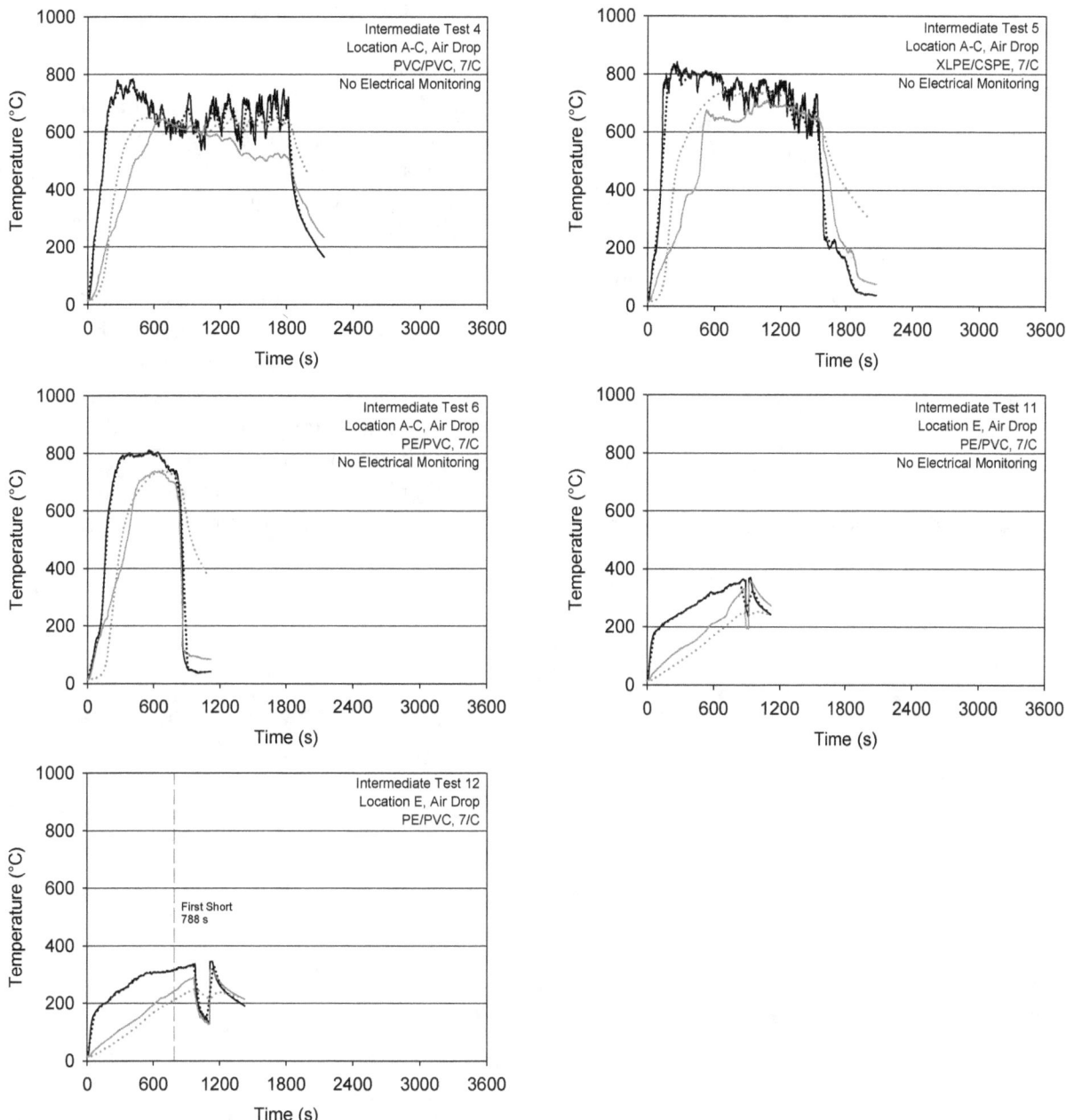

Figure 26. Summary of results for air drop cables in the Intermediate scale tests. Black indicates the exposing temperature; red the cable. Solid lines for experiment, dotted for model.

Six Cable Bundles

A common configuration for cables within the horizontal trays is shown below. Six cables were arranged in a small pyramid, two thermocouples were positioned above and below the bundle to measure hot gas temperatures, one thermocouple was positioned between two cables in the bundle, and one thermocouple was inserted under the jacket of one cable (labeled E in the figure). *Note that the letters associated with the cables within the bundle are not associated with the letters used to designate the various trays and conduits within the test rig.*

The results of the Intermediate Scale Tests involving 6 cable bundles are shown on the following pages. The bundles were laid in trays at Locations A, C, F, and G. Locations A, C, and F were within the fire plume, while G was not.

The THIEF model was tested using the measured gas temperature below the bundle (TC-3) as the *specified exposing temperature*. The predictions of the model were compared with the measured temperature within Cable E (TC-1). Where available, the graphs also display the measured electrical failure times of Cable E from the adjacent, electrically monitored bundle.

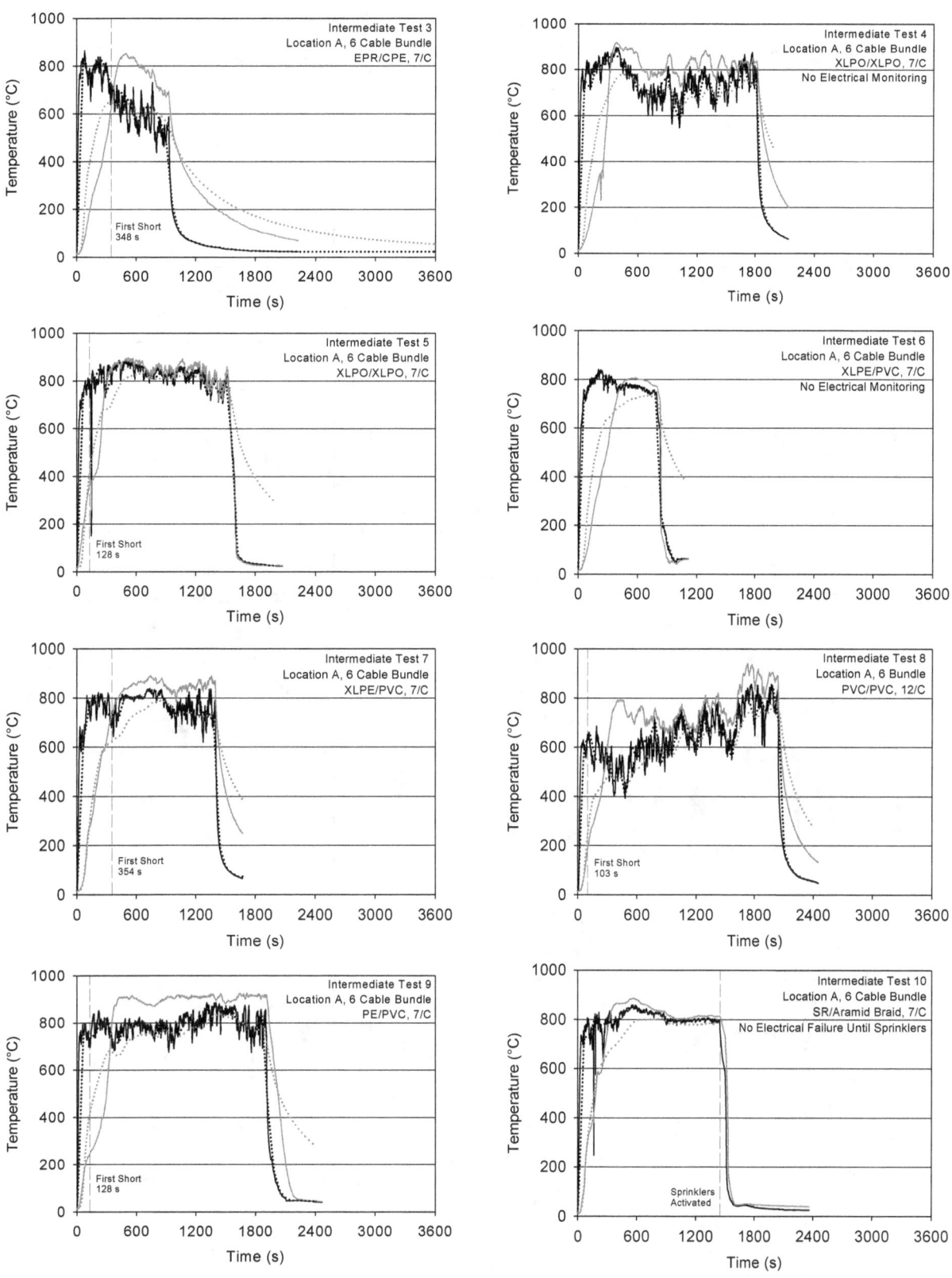

Figure 27. Summary of results for 6 cable bundles, Location A. Black indicates the exposing temperature; red the cable. Solid lines for experiment, dotted for model.

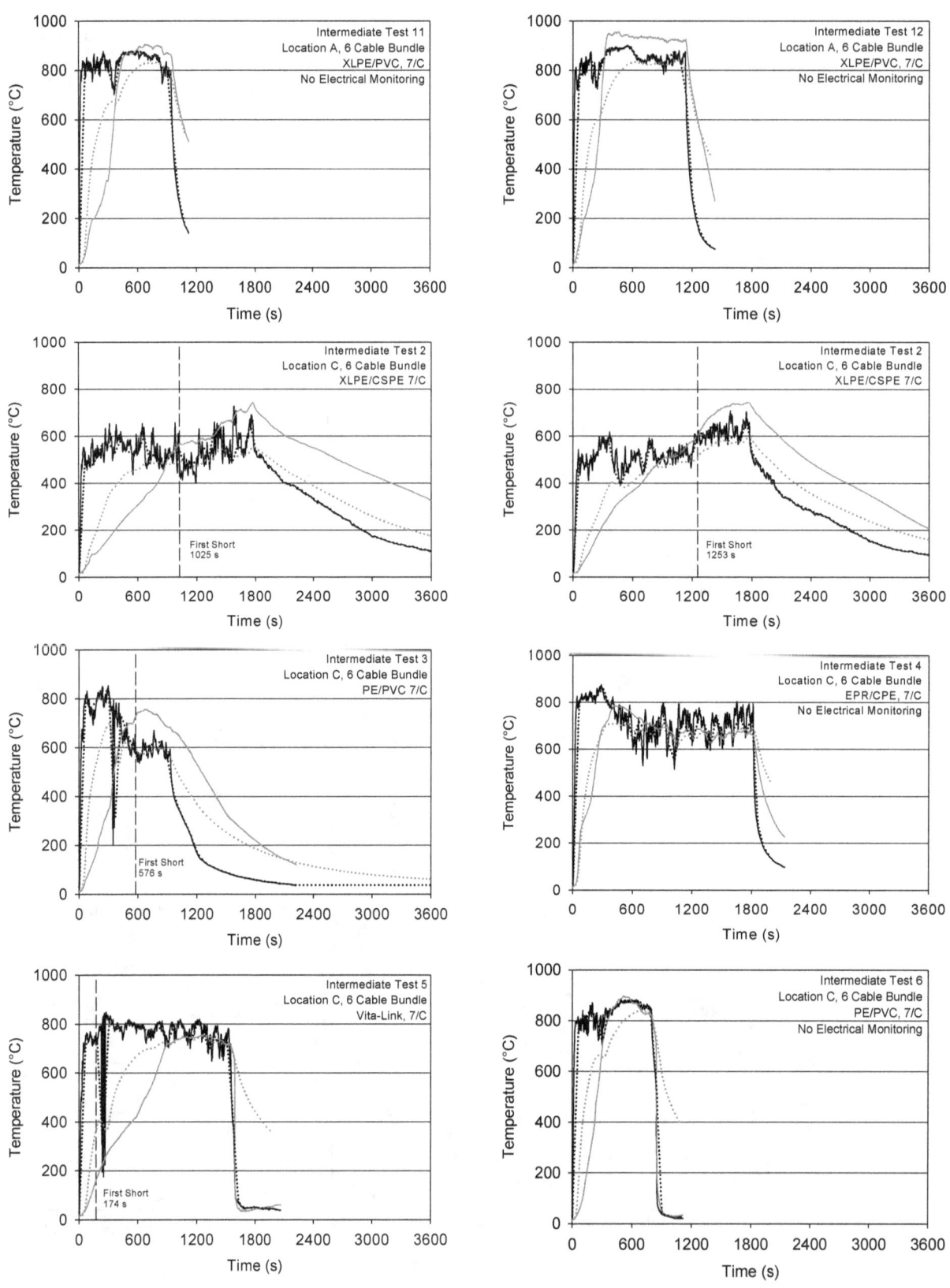

Figure 28. Summary of results for 6 cable bundles, Locations A and C. Black indicates the exposing temperature; red the cable. Solid lines for experiment, dotted for model.

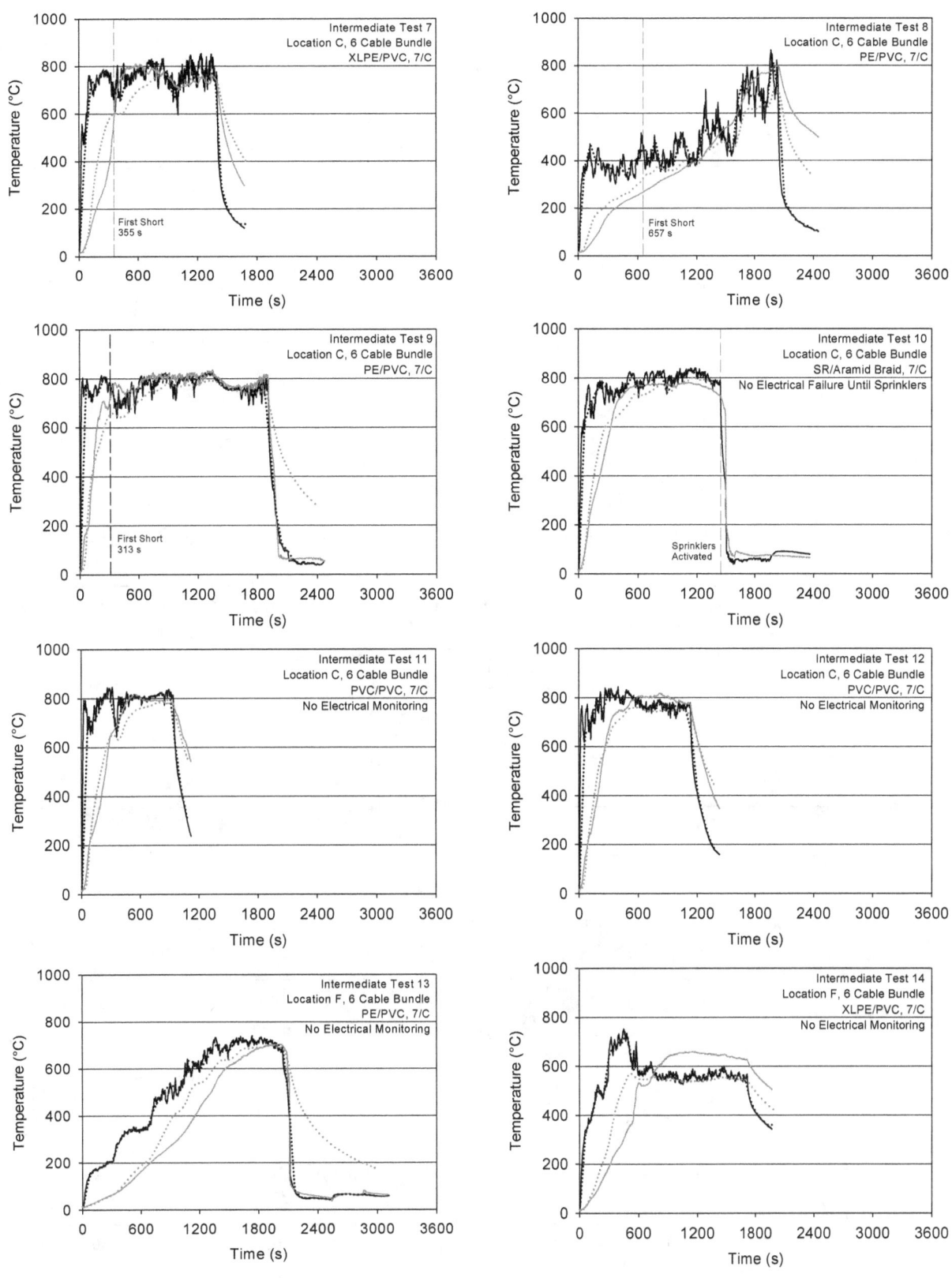

Figure 29. Summary of results for 6 cable bundles, Locations C and F. Black indicates the exposing temperature; red the cable. Solid lines for experiment, dotted for model.

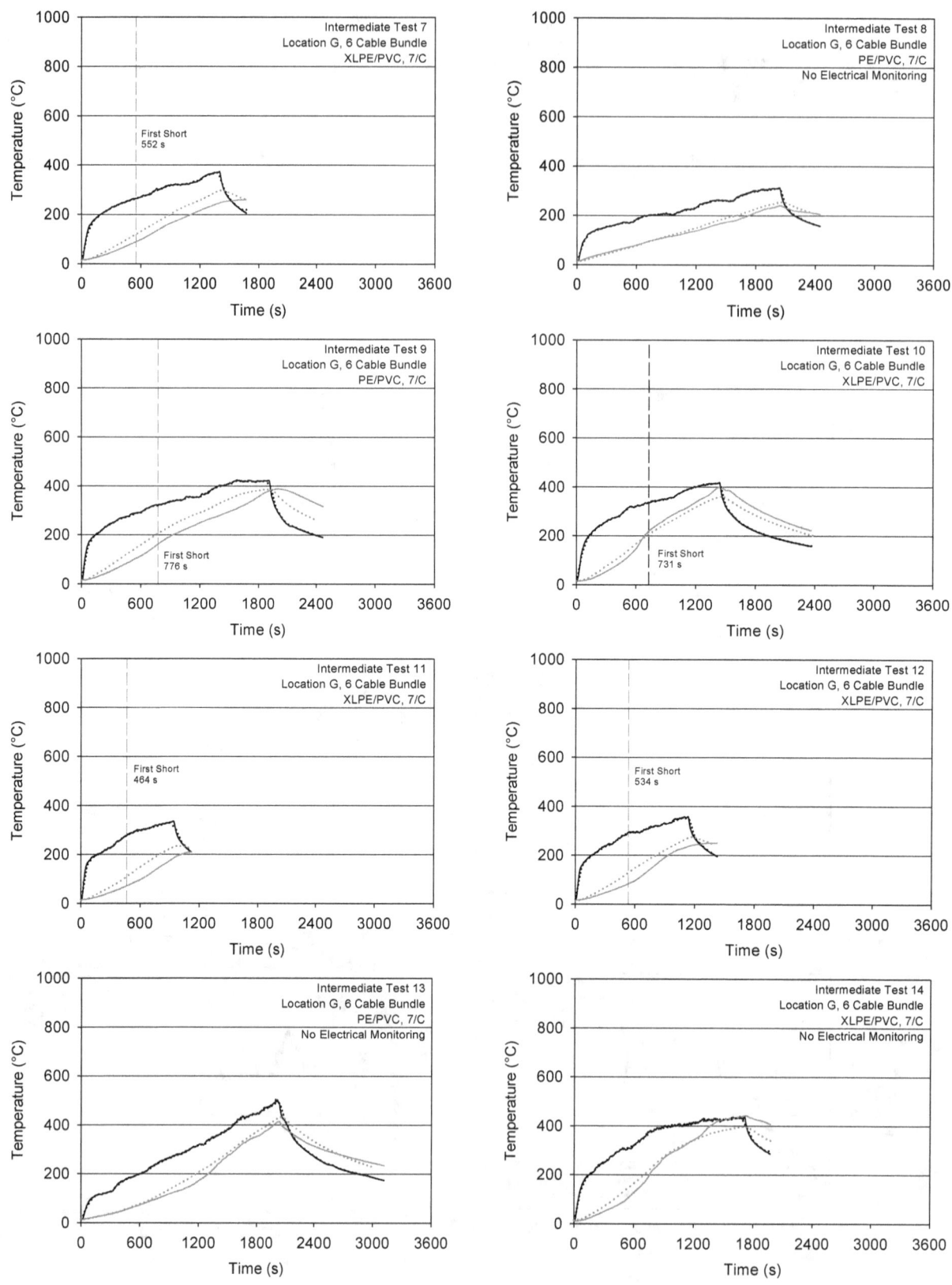

Figure 30. Summary of results for the 6 cable bundles, Location G. Black indicates the exposing temperature; red the cable. Solid lines for experiment, dotted for model.

Twelve Cable Bundles

In four instances during the Intermediate Scale Test Series, bundles of twelve cables were instrumented for thermal response, as shown below. Twelve cables were arranged in a small stack upon a ladder-backed tray, two thermocouples were positioned above and below the bundle to measure gas temperatures, one thermocouple was positioned between three cables in the bundle, and one thermocouple was inserted under the jacket of one cable (labeled A in the figure). *Note that the letters associated with the cables within the bundle are not associated with the letters used to designate the various trays and conduits within the test rig.*

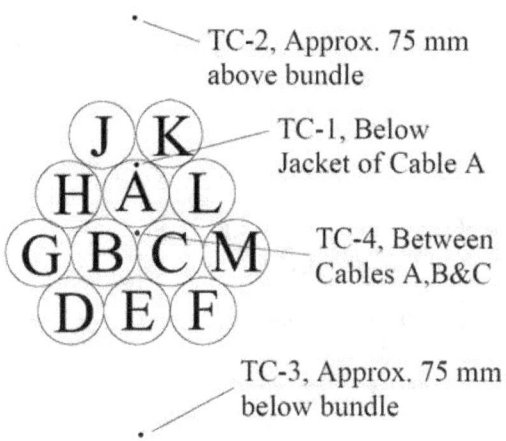

The results of the Intermediate Scale Tests involving 12 cable bundles are shown on the following page. The bundles were laid in trays at Locations A, B, and G. Location A was in the fire plume; Locations B and G were not.

The THIEF model was tested using the measured gas temperature above the bundle (TC-2) as the *specified exposing temperature*. The predictions of the model were compared with the measured temperature within Cable A (TC-1). Where available, the graphs also display the measured electrical failure times of Cable A from the adjacent, electrically-monitored bundle.

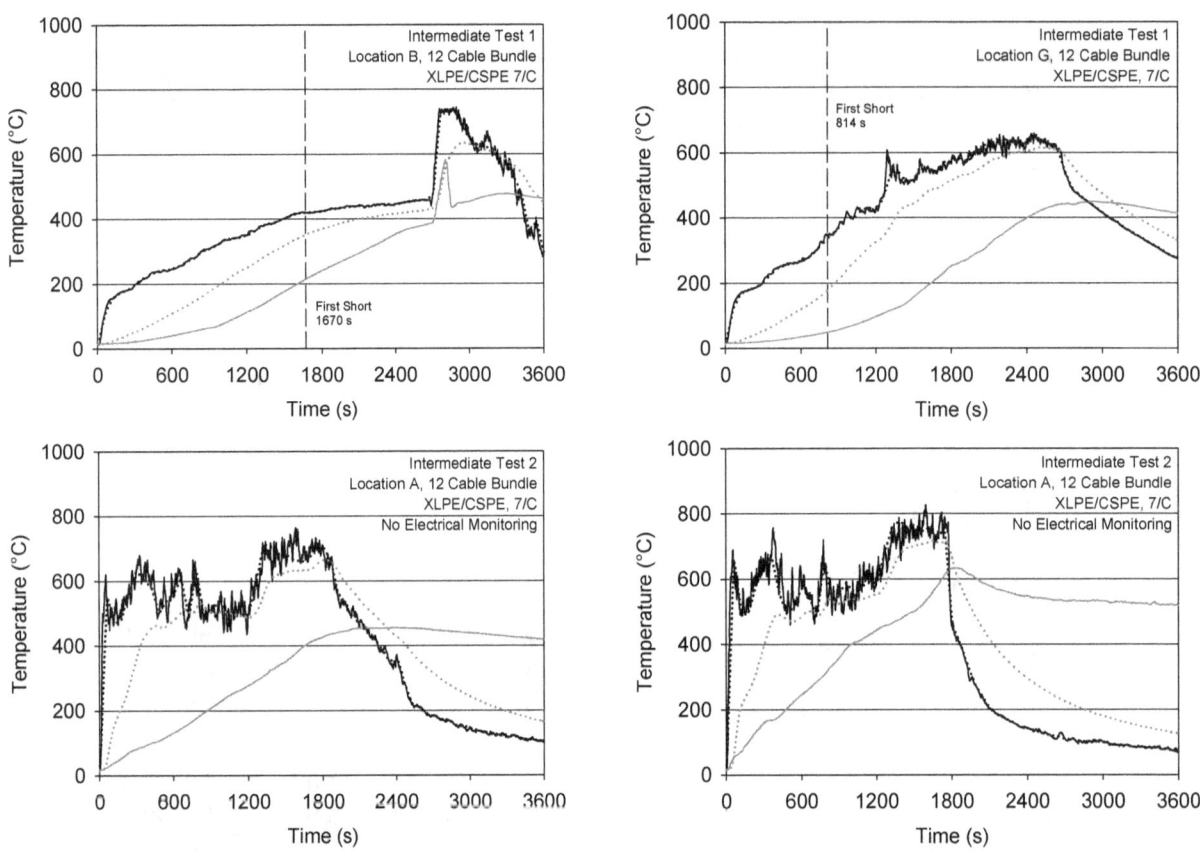

Figure 31. Summary of results for the 12 cable bundles. Black indicates the exposing temperature; red the cable. Solid lines for experiment, dotted for model.

4.4 Summary of the Intermediate Scale Tests

In all, 65 temperature predictions were compared to 65 measurements during the Intermediate Scale Test Series. The results are essentially 65 time histories of the predicted and measured inner cable temperatures. These graphs have been shown in the previous sections. To quantify the accuracy of the THIEF model, the same procedure that was developed for the Penlight results was followed. Instead of using the observed electrical failure times from the experiments, appropriate "threshold" values were used. The reason for this is that in the Intermediate Scale Test Series, some cables were monitored only for thermal response, some only for electrical response, and some were monitored for both using identical bundles. Consequently, it was not possible to draw conclusions about the THIEF model based on the measured failure times. A more appropriate test of the model is to compare its predictions of inner cable temperature directly with that which was measured for the time period between ignition of the fire and the point where the inner cable temperature measurements passed beyond a specified "threshold" value. From the Penlight series, the thermoset cables tested failed at temperatures between 400 °C and 450 °C (752 °F and 842 °F); the thermoplastics failed between 200 °C and 300 °C (392 °F and 482 °F). The exact failure temperatures are not particularly important for this exercise, rather the time to reach some threshold temperature consistent with the particular type of cable under consideration. For the Intermediate Scale Tests involving thermoset cables, 400 °C (752 °F) was chosen as the "threshold" temperature. For thermoplastics, 200 °C (392 °F) was chosen.

The results are shown in Figure 32 and tabulated in Table 4. For the 65 point to point comparisons, the THIEF model under-predicted the times to reach the "threshold" temperature by 15 %, on average, and the standard deviation was 33 %. The model predictions in this case are noticeably less accurate than the Penlight predictions. This is by design. The THIEF model was designed to under-predict cable failure times because it assumes that a given cable is completely exposed to the elevated temperatures of the surrounding hot gases. In reality, a cable is almost always shielded in some way by other cables, the tray, the conduit, and so on. Often cables are buried deep within a loaded tray of other cables and do not respond nearly as quickly to hot gases as the THIEF model would predict. Indeed, the most under-predicted failure times for the Intermediate Scale Test Series are those of the 12 cable bundles.

In predicting the outcomes of the Intermediate-Scale experiments, no attempt was made to modify or adjust the THIEF model to account for the relative position of the target cable within the bundle. Rather, the model was applied as it had been for the Penlight series, because that is the way the model is to be applied in practice. That is, it was assumed that the cable was not within a bundle, as there is no way to account for a bundle in the model. The reason for this assumption is that it is unlikely that a given cable randomly installed in a given tray will always be protected by its neighbors from hot gases of a fire. Thus, it is prudent to apply the THIEF model under the assumption that the cable will at some point along its length be directly exposed to the hot gases, and even if it is not, the prediction will err on the conservative side by predicting an early failure. Consider, for example, the points in Figure 32 labeled "6 Cable Bundle Inside Fire Plume." These refer to the scenarios where bundles containing six cables were placed in a tray positioned at location A or C, directly above the fire and within the flaming region. The THIEF model under-predicted the time to reach threshold for these points to a greater extent than it did for the six cable bundles outside the fire plume because the model assumes that each cable within the bundle is directly exposed to the very hot gases of the fire plume with no accounting for the protection offered by surrounding cables. Outside the fire plume, the gases are less hot and gradually penetrate the bundle over longer time periods. In effect, outside the fire plume, there is less of a difference in temperature between the gases outside and inside the bundle.

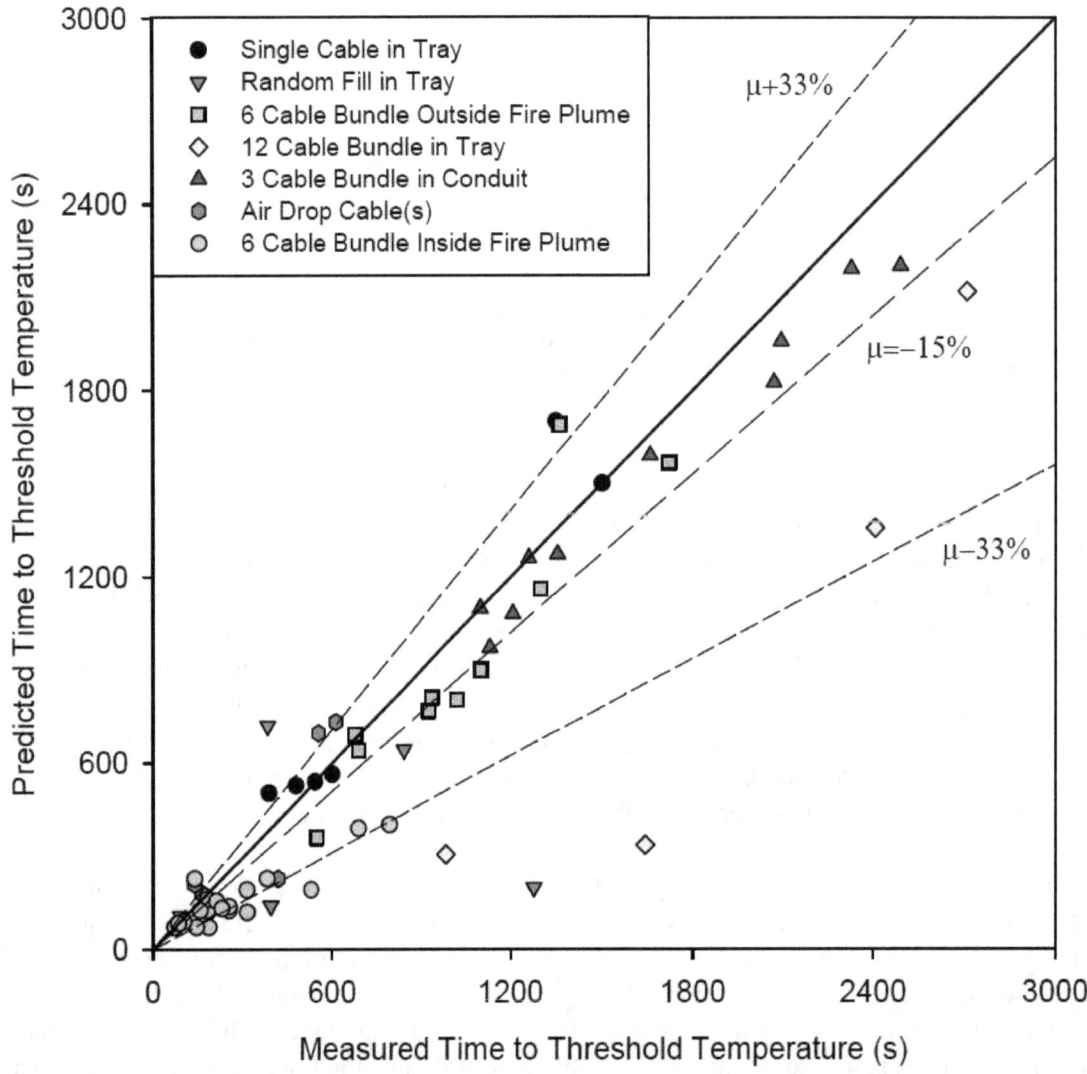

Figure 32. Summary of the Intermediate Scale Test predictions. The dashed lines indicate the average (-15 %) and standard deviation (33 %) of the data. Also note that the description "Inside Fire Plume" refers to locations A and C, which were within the flaming region of the fire.

Table 4. Results of the Intermediate Scale THIEF Model Predictions. See Nowlen and Wyant (2007b) for details about each configuration.

Test	Location Figure 17	Cable Configuration	Cable Composition (Insulation/Jacket)	Threshold Temperature (°C)	Measured Time to Threshold (s)	Predicted Time to Threshold (s)
1	A	Random Fill	XLPE/CSPE	400	390	140
1	A	Random Fill	XLPE/CSPE	400	380	720
1	B	12 Bundle	XLPE/CSPE	400	2710	2117
1	D	3+1 Conduit	XLPE/CSPE	400	2070	1824
1	E	3+1 Conduit	XLPE/CSPE	400	2490	2200
1	G	12 Bundle	XLPE/CSPE	400	2406	1355
2	A	12 Bundle	XLPE/CSPE	400	1644	336
2	A	12 Bundle	XLPE/CSPE	400	984	306
2	C	6 Bundle	XLPE/CSPE	400	792	402
2	C	6 Bundle	XLPE/CSPE	400	689	390
2	E	Single	PVC/PVC	200	540	540
2	G	Single	PVC/PVC	200	600	564
3	A	6 Bundle	EPR/CPE	400	252	126
3	C	6 Bundle	PE/PVC	200	185	72
3	E	Single	PVC/PVC	200	474	528
3	G	Single	PVC/PVC	200	384	504
4	A	6 Bundle	XLPO/XLPO	400	252	138
4	C	6 Bundle	EPR/CPE	400	185	126
4	A-C	Air Drop	PVC/PVC	200	162	180
4	E	Single	XLPE/CSPE	400	--	--
4	G	Single	XLPE/CSPE	400	--	--
5	A	6 Bundle	XLPO/XLPO	400	180	126
5	C	6 Bundle	Vitalink	400	527	192
5	A-C	Air Drop	XLPE/CSPE	400	414	228
5	E	Single	EPR/CPE	400	1350	1700
5	G	Single	EPR/CPE	400	1500	1500
6	A	6 Bundle	XLPE/PVC	400	138	228
6	C	6 Bundle	PE/PVC	200	144	72
6	A-C	Air Drop	PE/PVC	200	138	210
6	E	3 Conduit	PE/PVC	200	1098	1098
6	G	3 Conduit	PE/PVC	200	1260	1260
7	A	6 Bundle	XLPE/PVC	400	180	162
7	C	6 Bundle	XLPE/PVC	400	311	192
7	E	3+1 Conduit	PVC/PVC	200	1356	1272
7	G	6 Bundle	XLPE/PVC	400	1100	900
8	A	6 Bundle	PVC/PVC	200	102	90
8	C	6 Bundle	PE/PVC	200	378	228

Test	Location Figure 17	Cable Configuration	Cable Composition (Insulation/Jacket)	Threshold Temperature (°C)	Measured Time to Threshold (s)	Predicted Time to Threshold (s)
8	E	3+1 Conduit	XLPE/CSPE	400	2328	2190
8	G	6 Bundle	PE/PVC	200	1722	1566
9	A	6 Bundle	PE/PVC	200	72	72
9	C	6 Bundle	PE/PVC	200	84	84
9	E	3+1 Conduit	XLPE/CSPE	400	2094	1956
9	G	6 Bundle	PE/PVC	200	924	768
10	A	6 Bundle	SR/Aramid	400	156	126
10	C	6 Bundle	SR/Aramid	400	210	156
10	E	3+1 Conduit	PE/PVC	200	1206	1080
10	G	6 Bundle	XLPE/PVC	400	678	690
11	A	6 Bundle	XLPE/PVC	400	312	120
11	C	6 Bundle	PVC/PVC	200	72	72
11	E	3+1 Air Drop	PE/PVC	200	552	696
11	G	6 Bundle	XLPE/PVC	400	1020	804
12	A	6 Bundle	XLPE/PVC	400	228	132
12	C	6 Bundle	PVC/PVC	200	84	84
12	E	3+1 Air Drop	PE/PVC	200	612	732
12	G	6 Bundle	XLPE/PVC	400	936	810
13	A	Random Fill	XLPE/CSPE	400	1278	198
13	C	Random Fill	XLPE/CSPE	400	840	642
13	E	3+1 Conduit	XLPE/CSPE	400	1660	1590
13	F	6 Bundle	PE/PVC	200	690	640
13	G	6 Bundle	PE/PVC	200	1300	1160
14	A	Random Fill	XLPE/CSPE	400	90	108
14	C	Random Fill	XLPE/CSPE	400	180	162
14	E	3+1 Conduit	PE/PVC	200	1130	970
14	F	6 Bundle	XLPE/PVC	400	546	360
14	G	6 Bundle	XLPE/PVC	400	1362	1690

5. LESSONS LEARNED FROM CAROLFIRE

Both the Penlight and Intermediate scale experiments and simulations discussed in the previous sections raise several issues that affect the practical implementation of the THIEF algorithm into a fire model.

5.1 Characterizing the thermal environment

The boundary condition for the THIEF model is the net heat flux to the cable surface. If the cable or cables are within a hot, smoky gas layer, then the heat flux can be calculated in terms of the surrounding gas temperature, T_g :

$$\dot{q}'' = \varepsilon\sigma\left(T_g^4 - T_s^4\right) + h\left(T_g - T_s\right) \tag{5.1}$$

The gas temperature is computed by the fire model, and is usually a function of time. The convective heat transfer coefficient, h, is usually assumed to be constant, except in computational fluid dynamics models, where it can be calculated based on local flow conditions. The cable surface temperature, T_s, is computed within the THIEF algorithm.

If the cable or cables are directly exposed to the fire, Eq. (5.1) does not apply, as then the heat flux is not simply a function of the surrounding gas temperature. Different types of fire models treat this situation differently, but regardless, the net heat flux to the cable is:

$$\dot{q}'' = \varepsilon\dot{q}_{\text{inc}} - \varepsilon\sigma T_s^4 \tag{5.2}$$

where the subscript "inc" refers to the *incident* heat flux. A point source radiation model, for example, calculates the incident heat flux, from which the *net* heat flux can be obtained by subtracting off the re-radiated component.

5.2 Specifying a "failure temperature"

The THIEF model only predicts the interior temperature of a cable. It is assumed by the user that the cable fails at some experimentally determined temperature. For most of the cables tested in the CAROLFIRE program, electrical failure correlated fairly well to interior temperature, with thermoplastic cables failing at temperatures between 200 °C and 250 °C (392 °F and 482 °F); and thermosets failing between 400 °C and 450 °C (752 °F to 842 °F) (Nowlen and Wyant 2007b). However, there were several cables tested that did not fail at these temperatures. In fact, several cables ignited and burned, but did not fail electrically until water was applied.

5.3 Defining the cable within the fire model

THIEF is only a *target* model, like the activation model of a sprinkler or smoke detector. It requires that the fire model in which it is embedded as a subroutine provide it only with a heat flux as input and it only outputs a "failure time" – it does not affect the overall thermal environment unless it is explicitly included in the fire model as an object that drains energy from the surrounding gases, like a wall or larger obstruction. Most simple fire models do not explicitly include targets as objects that can affect the overall thermal environment. In fact, the THIEF calculation could be performed separately as part of the post-processing phase.

5.4 Modeling cable burning

It was noted by Nowlen and Wyant (2007b) during the CAROLFIRE experiments that cable ignition often occurred just after electrical failure. It was speculated that the short circuit acts like a pilot to ignite flammable vapors that off-gas as the cables heat up. The THIEF model can predict the temperature rise within the cable to a level of accuracy that has been demonstrated in the previous two sections. However, the model cannot predict ignition and burning. This is not to say that the fire model within which THIEF is implemented cannot predict ignition and burning; but this depends on the type of model. For example, empirical correlations, like those described in NUREG-1805 (Iqbal and Salley 2004), use experimentally obtained ignition temperatures and burning rates to estimate the heat release and spread rates of various types of cables. More detailed fire models can predict, to some degree, ignition and burning, but these models require far more thermo-physical property data than does THIEF, and this still remains a critical hurdle in developing this functionality in the models.

6   CONCLUSION

A thermally-induced electrical failure (THIEF) model for cables has been shown to work effectively in realistic fire environments. The THIEF model is essentially nothing more than the numerical solution of the one-dimensional heat conduction equation within a homogenous cylinder with fixed, temperature-independent properties. The model was used to predict the inner cable temperature of 100 instrumented cables from the CAROLFIRE Penlight (35 single cable experiments; 66 point to point comparisons) and Intermediate Scale Test Series (14 experiments; 65 point to point comparisons). Because the Penlight experiments tested single cables that were heated uniformly on all sides, the one-dimensional THIEF model accurately predicted the times for the temperature inside the cable jacket to reach "threshold" values that are typically observed when the cable fails electrically. For 66 measurements, the model under-predicted the time to reach threshold temperature by 3 %, on average. In the Intermediate Scale experiments, where the cable configurations were more typical of actual installations, the model under-predicted the times to reach threshold temperature by 15 %, on average. This latter result is realistically conservative – the THIEF model does not account for the shielding effects of cable bundles, and thus over-predicts cable temperatures and under-predicts "failure" times.

The cables included in the study ranged from 7 mm (0.25 in) to 19 mm (0.75 in) in diameter, a common size for control cables, plus some instrument and low power cables. The copper content by volume ranged from 0.07 to 0.36 and the content by mass ranged from 0.31 to 0.89. The volume and mass fractions are not direct model inputs, but rather the average density of the cable as a whole. Nevertheless, the range in cable properties demonstrates that the THIEF model is applicable to a wide variety of cables with no need for additional information beyond the cable diameter, mass per length, and an empirical "failure" temperature. In addition, there was no indication from the model results that indicated a bias related to the number of conductors, plastic composition, or copper content.

While there are various ways to refine the THIEF model – multiple layers of materials, two or three spatial dimensions, temperature dependent thermal properties, polymeric decomposition, and so on – it is unlikely that any of these enhancements would dramatically improve its overall accuracy, especially in light of the uncertainty associated with the fire simulation of the entire compartment. According to a recent NRC/EPRI verification and validation (V&V) study of five different fire models (NUREG-1824/EPRI 1011999), the error in the predicted net convective and radiative heat transfer to various "targets" is on the order of 20 % or higher, depending on the type of model. Given an uncertainty of 20 % in the exposing heat flux for the simple cable failure calculation, it is unclear how additional complexity would generate better results than those presented in this report.

7 REFERENCES

1. Andersson, P. and P. Van Hees (2005) "Performance of Cables Subjected to Elevated Temperatures," *Fire Safety Science – Proceedings of the Eighth International Symposium*, International Association for Fire Safety Science.

2. Chourio, G. (2007). *Probabilistic Models to Estimate Fire-Induced Cable Damage in Nuclear Power Plants*, Ph.D. Thesis, University of Maryland, College Park.

3. Hamins, A, A. Maranghides, R. Johnsson, M. Donnelly, J. Yang, G. Mulholland and R.L. Anleitner (2006). *Report of Experimental Results for the International Fire Model Benchmarking and Validation Exercise #3*, NIST Special Publication 1013-1, National Institute of Standards and Technology, Gaithersburg, Maryland (Also published by the US Nuclear Regulatory Commission as NUREG/CR-6905).

4. Incropera, F.P. and D.P. DeWitt (1990). *Fundamentals of Mass and Heat Transfer*, (3rd ed.), John Wiley & Sons, New York.

5. Iqbal, N. and M. Salley (2004). *Fire Dynamics Tools*, NUREG-1805, U.S. Nuclear Regulatory Commission, Washington, DC.

6. McGrattan, K.B, S. Hostikka, J.E. Floyd, H.R. Baum and R.G. Rehm (2007). *Fire Dynamics Simulator (Version 5) Technical Reference Guide*, NIST Special Publication 1018-5, National Institute of Standards and Technology, Gaithersburg, Maryland.

7. Nowlen, S.P. and Wyant, F.J. (2007a). *CAROLFIRE Test Report Volume 1: General Test Descriptions and the Analysis of Circuit Response Data*, NUREG/CR-6931/V1, US Nuclear Regulatory Commission, Washington, DC.

8. Nowlen, S.P. and Wyant, F.J. (2007b). *CAROLFIRE Test Report Volume 2: Cable Fire Response Data for Fire Model Improvement*, NUREG/CR-6931/V2, US Nuclear Regulatory Commission, Washington, DC.

9. U.S. NRC (1975). "Cable Fire at Browns Ferry Nuclear Power Station," NRC Bulletin BL-75-04, U.S. Nuclear Regulatory Commission, Washington, DC, March.

10. U.S. NRC and EPRI (2007). *Verification and Validation of Selected Fire Models for Nuclear Power Plant Applications*, NUREG-1824, U.S. Nuclear Regulatory Commission, Rockville, Maryland.

11. Weast, R.C. (ed.). (1982) *CRC Handbook of Chemistry and Physics*, CRC Press, Inc., Boca Raton, Florida.

NRC FORM 335 (9-2004) NRCMD3.7	U.S. NUCLEAR REGULATORY COMMISSION	1. REPORT NUMBER (Assigned by NRC, Add Vol., Supp., Rev., and Addendum Numbers, if any.)
BIBLIOGRAPHIC DATA SHEET *(See instructions on the reverse)*		NUREG/CR-6931, Vol. 3

2. TITLE AND SUBTITLE

Cable Response to Live Fire (CAROLFIRE), Volume 3: Thermally-Induced Electrical Failure (THIEF) Model

3. DATE REPORT PUBLISHED

April	2008

4. FIN OR GRANT NUMBER

N6414

5. AUTHOR(S)

Kevin McGrattan

6. TYPE OF REPORT

Technical

7. PERIOD COVERED *(Inclusive Dates)*

8. PERFORMING ORGANIZATION – NAME AND ADDRESS *(If NRC, provide Division, Office or Region, U.S. Nuclear Regulatory Commission, and mailing address; if contractor, provide name and mailing address.)*

National Institute of Standards and Technology
Building and Fire Research Laboratory
Gaithersburg, Maryland 20899-8663

9. SPONSORING ORGANIZATION – NAME AND ADDRESS *(If NRC, type "Same as above"; if contractor, provide NRC Division, Office or Region, U.S. Nuclear Regulatory Commission, and mailing address.)*

Division of Risk Analysis
Office of Nuclear Regulatory Research
U.S. Nuclear Regulatory Commission
Washington, DC 20555-0001

10. SUPPLEMENTARY NOTES

J. Dreisbach, NRC Project Manager

11. ABSTRACT *(200 words or less)*

This report describes a thermally-induced electrical failure (THIEF) model's ability to predict the behavior of power, instrument, and control cables during a fire. The model is intended to be incorporated as a subroutine for deterministic fire models, and it is of comparable accuracy and simplicity to the activation algorithms for various other fire protection devices (e.g., sprinklers, heat and smoke detectors). THIEF model predictions are compared to experimental measurements of instrumented cables in a variety of configurations, and the results indicate that the model is an appropriate analysis tool for nuclear power plant applications. This work was performed as part of the CAROLFIRE (Cable Response to Live Fire) program sponsored by the U.S. Nuclear Regulatory Commission. The experiments for CAROLFIRE were conducted at Sandia National Laboratories, Albuquerque, New Mexico. Details of the CAROLFIRE experimental program are contained in Volumes 1 and 2 of this three-volume series.

12. KEY WORDS/DESCRIPTORS *(List words or phrases that will assist researchers in locating the report.)*

Fire Modeling
Risk-Informed

13. AVAILABILITY STATEMENT

unlimited

14. SECURITY CLASSIFICATION

(This page)
unclassified

(This report)
unclassified

15. NUMBER OF PAGES

16. PRICE

NRC FORM 335 (9-2004)

www.ingramcontent.com/pod-product-compliance
Lightning Source LLC
Chambersburg PA
CBHW081841170526
45167CB00007B/2874